# Biological Rhythms and
# Human Performance

# Biological Rhythms and Human Performance

Edited by

## W. P. COLQUHOUN

*MRC Unit of Applied Psychology,*
*Laboratory of Experimental*
*Psychology, University of Sussex,*
*Brighton, Sussex, England*

1971 (AP) ACADEMIC PRESS · LONDON
AND NEW YORK

ACADEMIC PRESS INC. (LONDON) LTD
24-28 Oval Road
London NW1 7DX

*U.S. Edition published by*
ACADEMIC PRESS INC.
111 Fifth Avenue,
New York, New York 10003

Library of Congress Catalog Card Number: 75–153524
ISBN: 0–12–182050–5

Printed by photolithography in Great Britain by
T. & A. Constable Ltd., Edinburgh

# Contributors

M. J. F. BLAKE,† *MRC Applied Psychology Unit, Cambridge, England*

W. P. COLQUHOUN, *MRC Unit of Applied Psychology, Laboratory of Experimental Psychology, University of Sussex, Brighton, Sussex, England*

B. C. GOODWIN, *School of Biological Sciences, University of Sussex, Brighton, Sussex, England*

K. F. H. MURRELL, *Applied Psychology Department, University of Wales Institute of Science and Technology, Cardiff, Wales*

KEITH OATLEY, *Laboratory of Experimental Psychology, University of Sussex, Brighton, Sussex, England*

JUNE A. REDGROVE, *Department of Part-Time Business Studies and Management, Enfield College of Technology, Enfield, Middlesex, England*

A. J. SANFORD, *Department of Psychology, The University, Dundee, Scotland*

WILSE B. WEBB, *Department of Psychology, University of Florida, Gainesville, Florida, U.S.A.*

† Deceased.

# Preface

It has long been known that most, if not all, biological processes are rhythmic or cyclical in character, that is, they consist of sequences of events which repeat themselves at regular intervals. The sequence may be very simple or extremely complex, and the time-span, or period of a particular cycle may be anything from a fraction of a second to many years. Such rhythms occur in all forms of life, and man is no exception.

In this volume an attempt has been made to illustrate some of the ways in which certain characteristic human rhythms influence behaviour through their effects on the functioning of the nervous system. Particular emphasis has been laid on measures of actual performance rather than on feelings or mood, not because rhythms in the latter are any less apparent, but because emotional changes are far more difficult to quantify than perceptual, motor and cognitive functions which can be assessed by objective tests.

After an introductory chapter by Drs. Oatley and Goodwin on the general nature of biological rhythms and the methods most commonly employed in their measurement and analysis, a major section of the book is devoted to what is probably the best known rhythm of all—the diurnal, or 24-h activity cycle around which the greater part of our daily life is organized. The Editor reviews the literature on the effects of this rhythm on mental efficiency, and describes some recent large-scale experiments designed to assess the significance of these effects in the round-the-clock operation of plant and machinery by teams of shift-workers. This is followed by an account of some hitherto unpublished research by Mr. Blake into the question of individual differences in diurnal performance rhythms and their relation to personality and physiological factors. Professor Webb then discusses the phenomenon of sleep and its cyclic structure, considering the sleep process as a "performance" which is itself subject to circadian variation like so many other functions of the organism.

The next two chapters demonstrate the wide range of time-scales over which biological rhythms influence performance. As an example of the effects of very short cycles, Dr. Sanford examines the relationship between the well-known alpha rhythm of electrical brain activity and the timing of perceptual and motor activities; he then goes on to discuss the more general question of periodicity in the processing of information by the nervous system. At the other end of the scale, Dr. Redgrove is concerned with the ways in which the various phases of the menstrual cycle may affect the efficiency with which female operatives carry out their day-to-day work in office or factory.

Finally, in a chapter entitled "Industrial Work Rhythms", Professor Murrell tackles the problems of measuring and accounting for the fluctuations in production which often occur on the actual shop-floor from hour to hour or from day to day. The research described here highlights the difficulties of distinguishing between the effects of biological rhythms and those arising from social habits or other factors.

Each of the authors of the different chapters of this volume is an expert in the field which he or she describes, and each has carried out original research in that area. Although in a work of this size it is not possible to cover the entire range of rhythmic processes which might influence behaviour, it is believed that the investigations described by the various contributors cover all those major sources of cyclical variation in human performance about which any considerable body of knowledge exists at the present time. Thus it is hoped that this book will serve both as a comprehensive review of what is actually known about these rhythmic phenomena, and as a stimulus for future research in those fields where the extent of our understanding is still considerably limited.

*W. P. COLQUHOUN*
*August, 1971*

# Contents

# 7. Industrial Work Rhythms

K. F. H. MURRELL

CHAPTER 1

# The Explanation and Investigation of Biological Rhythms

KEITH OATLEY

*Laboratory of Experimental Psychology, University of Sussex, Brighton, Sussex, England*

and

B. C. GOODWIN

*School of Biological Sciences, University of Sussex, Brighton, Sussex, England*

## I. Introduction

The rhythmic occurrence of events, though commonplace in biology, tends to be overlooked and has not until relatively recently been considered as fundamental to living systems. Though, for instance, growth and division of cells is inherently periodic, neither periodicity itself nor the oscillatory nature of some of the control processes involved have always figured largely in explanations of the events in question. Instead experimenters have sought to stabilize the system they are investigating and to pay attention to the effect of a stimulus on conceptually rather passive systems.

This same situation tends to exist in psychology and the study of human performance. Even though rhythmicitics are obvious if they are sought, they have not attracted much attention. Supposedly static conditions have been studied, helped by a mixture of standardized stimulus arrangements and statistical tests. However, if there are important periodic processes within an experimental subject, a stimulus at one time will not have the same effect as the same stimulus at another time, and carrying out statistical tests on results merely assumes that any variations are noise. An alternative is that periodic variations are of the essence.

Rhythms provide clues as to the existence of oscillators which have important properties about which many people are ignorant. One such property of a class of oscillators known as non-linear (defined in the next section) is that of entrainment. Huygens, who lived in the 18th century, seems to have been the first to notice this (Minorsky, 1962). He observed that when 2 clocks which separately ran at different speeds were both hung on the same thin wooden board, they became synchronized and kept the same time. Evidently very feeble interactions due to the ticking being transmitted from one clock to the other through the board were capable of synchronizing the 2 clocks together. The tendency to become entrained in this way by even very weak signals, so long as they are periodic, is a perfectly general property of non-linear oscillators; of which the clock is one example and most biological rhythm generators are others. The mere presence of this and other such properties of oscillators throughout biological systems demands a certain amount of rethinking of some traditional concepts.

Rhythms which may be important in performance range from high frequencies of the order of 10 cyc/sec (Hz) for possible periodicities underlying the intake of stimulus events and the release of motor patterns (see Chapter 5) to the low frequency, menstrual rhythms the existence of which is rather better substantiated (see Chapter 6). Some of these rhythmic processes may be fundamental to performance. If for instance skilled tracking can only be performed by release of discrete ballistic movements in a periodic sequence (Wilde and Westcott, 1962), then the periodicity may be basic. In other circumstances, the rhythm may not be part of the performance process itself, and understanding it therefore will have no very profound effect on our understanding of the mechanisms underlying performance. This may be the case for instance in the effect of the menstrual cycle. Nevertheless whether rhythmicity is a fundamental or merely an interfering variable, in many applied situations it is clear that optimal performance or optimal experimental control will only be attained by recognizing rhythms, and co-operating with them.

After defining some terms we will examine the essential nature of biological rhythms, from a very general point of view. Then the biological function of some rhythms that affect human performance will be discussed. This is followed by an account of a few special methods used to identify rhythmicity, and to investigate the basis of rhythms.

## II. Terms and Definitions

Several terms of importance are defined here in order to clarify later discussion. Rhythms are produced by oscillators. Oscillations can be divided into 2 main types, linear and non-linear. A linear oscillation is a sine wave (see Fig. 1A), or pure frequency; the oscillation of simple harmonic motion. Sinusoidal motion is exhibited by pendulums, tuning forks, and various electronic circuits containing a capacitor and an inductor, and can be described by a second-order linear differential equation. It is produced when 2 energy stores transfer energy alternatively from one to the other (e.g. kinetic and potential energy in a pendulum) at rates which

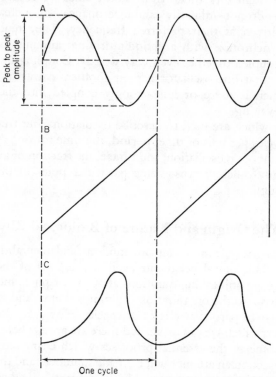

FIG. 1. Two cycles each of (A) a linear, sinusoidal oscillation; (B) a non-linear relaxation oscillation; and (C) a non-linear limit cycle oscillation. One cycle is equivalent to 360° or $2\pi$ radians of phase angle.

depend linearly on the amount stored in each at every instant of time.

Non-linear oscillations have a waveform which differs from a sinusoid (see Fig. 1B). A common non-linear form is a relaxation oscillation. Here a steadily rising (or falling) variable crosses a threshold, and then resets to a previous value. The sawtooth waveform of an oscilloscope timebase is a relaxation oscillation, so is a bank balance, draining away little by little and being restored (more or less) at monthly intervals. A second, and slightly more complex type of non-linear oscillation is the limit-cycle type (Fig. 1C). Plotting amplitude against its rate of change, a closed but not circular curve can be drawn and displacements of the oscillation away from this curve are followed by an asymptotic return to this same trajectory. Again, a differential equation (of at least second order) is required to describe the system, but the equation is a non-linear one. For instance, the rate of transfer of energy from one store to another may not be linearly proportional to the amount stored.

Non-linear oscillators differ from linear ones in several important respects. They do not ordinarily exhibit resonance—an increase in amplitude when driven at their preferred frequency. They tend instead to entrain, or synchronize with a periodic driving stimulus, or with each other when coupled together even very weakly. Coupling in this sense is any influence that one oscillator has on another. An alarm clock is an oscillator which is more or less strongly coupled into the rhythm of sleeping and waking.

Parameters which are used to describe oscillations are: **frequency,** the number of cycles per unit of time; **period,** the time for one cycle; **amplitude,** the extent of an oscillation; and **phase,** the fraction of a whole cycle, or in a slightly different sense some particular point in the cycle, e.g. half way through.

## III.   The Origin and Nature of Biological Rhythms

We are a long way from having any kind of fundamental theory for the explanation of biological behaviour even at the level of the cell or the organism. We are still asking questions about what phenomena are to be regarded as basic to the organization of living systems, and what as adaptations peculiar to particular classes or species. However, there are some generalizations which we can make, and there are areas where it is possible to describe some of the essential processes which characterize the living state. In this section an attempt will be made to understand the occurrence of rhythmic activity in organisms within the context of the overall strategy of the living process, conceived as the perpetuation of a state of stable dynamic organization. The individual contributions to this volume may then be seen as detailed probings into selected aspects of the dynamics of this process, attempting to assess the functional significance and the dynamic origins of various observed rhythms. In such work one is constantly confronted with the questions: how does a particular physiological rhythm contribute to the total stability, the adaptive performance, of the organism, and what is the origin of the oscillatory activity? Very general answers to these questions will now be attempted as they arise in relation to the most elementary and basic aspects of the living process, in the hope that these considerations may be relevant to an understanding of the more intricate dynamic behaviour of complex organisms.

A. *Oscillatory behaviour in open systems: rhythms of intrinsic or extrinsic origin*

The living organism is an open system with matter and energy flowing across its boundaries. There are no known quantitative laws which can be used to characterize and define the steady states of such systems, as there are for the characterization of equilibrium states in closed chemical or biochemical systems, although progress in this direction is certainly being made (Prigogine, 1967). These latter laws are those of thermodynamics which tell us, for example, that the equilibrium state is defined by the condition of maximum entropy and minimum free energy. It is of some interest to observe that continuous rhythmic or oscillatory activity in system variables is forbidden by thermodynamic laws for closed systems. If these laws applied to the living organism, then rhythmic activity would be excluded from their behaviour. However, being open systems, living organisms can occupy states considerably displaced from thermodynamic equilibrium. One of the prominent dynamical features of these states is the occurrence of rhythmic activity. Since there is as yet no rigorous physico-chemical explanation and analysis of this behaviour, it is necessary to use biological arguments in attempting to understand its origin and significance.

There are two essentially different possible explanations of rhythmic activity in biological systems. The first is that it is an essential dynamical feature of the process observed. The second is that such rhythms represent adaptive responses of the organism to a periodic environment. If the organism were not or had not been exposed to the periodicity, it would not show the rhythm. We will give examples of each. It is not always easy to decide which explanation is valid for any particular observed rhythm.

One of the fundamental periodicities in living systems is the cycle of growth and division in single cells. Elementary considerations about this process lead to the conclusion that there must be a dynamic oscillation in one at least of the state variables characterizing the system in order for the cycle to occur. The argument is as follows. It is known that the deoxyribonucleic acid (DNA) of the cell starts to be replicated at a particular time in the overall cycle, a time which is well defined in relation to some reference point such as the moment of daughter cell separation. Hence there must be a "clock" of some kind, an oscillator which completes its circuit once every cell cycle and provides the phase information necessary for the initiation of DNA replication at a defined point in the cycle. Such an oscillator is essential to the process. It does not arise from any environmental periodicity.

Having established that the cycle of growth and division of single cells as it is observed necessarily involves the occurrence of a biochemical rhythm or oscillation in the cell, it is still important to enquire why this particular strategy of organization has been adopted by cells. For it is conceivable that growth and division should not involve an internal clock of the type described. We could imagine, for example, that all cellular constituents are made at rates which maintain constant ratios between concentrations, and that when a critical cell size is reached an instability arises which results in cell division. Such a pattern of uniform, homogeneous growth with a rather sudden instability of mass would not involve any biochemical rhythm within the system, since all constituent molecular species would be synthesized at uniform rates and maintain fixed ratios. The reason why such a strategy cannot be followed by living cells as we know them is elementary but rather important. It is absolutely essential that certain processes be completed before others begin in the growth cycle. For example, the replication of the hereditary material, the DNA, must be completed before the cell divides in two, otherwise one of the daughter cells would be deficient in some genes and would be unable to survive. The control circuit which regulates the initiation of DNA replication must operate in such a manner that at some point in the cell reproductive cycle DNA replication is initiated; and at some other point this process must be terminated. These events must be well ordered in relation to the event of daughter cell separation, to guarantee that both daughters receive their full set of parental genes. The intracellular signals for initiation of DNA replication and for cell division thus occur in a well-defined temporal order. And since these signals are repeated in every cell cycle each time the system reproduces itself, we see immediately that the whole process is intrinscially rhythmic or oscillatory in character. Temporal constraints of this kind require oscillations to provide the phase information necessary to achieve the requisite temporal organization of the system (Goodwin, 1969).

This argument generalizes to any system in which such constraints operate within an overall repetitive or cyclic process. For example, in the menstrual cycle the fact that a discrete event, the release of the egg from the ovary, is basic to the whole process is sufficient to necessitate the occurrence of at least one physiological signal in the process, that controlling ovulation. Since this signal and its related event must be repeated once on every cycle, there must be at least one physiological oscillation. Of course, in this particular cycle there are several oscillations or rhythms, all bearing well-defined phase relations to one another. In fact, it would appear that whenever a basic process requires an oscillation for its repeated realization, the events connected with that process also become temporally ordered in a manner which is maximally adaptive for the process as a whole. Thus the number of constituent rhythmic activities increases. These secondary rhythms may become autonomous in the sense that they may continue in the absence of the presumed primary rhythm, as appears to occur in the case of cell division in bacteria, since it has been observed that there are mutants which cannot replicate their DNA but which continue to grow and divide. Alternatively, secondary rhythms may be driven by the primary oscillator. Or it may happen that the total set of interconnected rhythmic processes becomes the necessary and sufficient condition for the overall cycle, none being autonomous and the whole cycle failing if any one is arrested.

These considerations establish the fact that oscillations may be expected in biological systems whenever temporal constraints are required to separate incompatible events in a process. Such oscillations will have frequencies determined by the time required for completion of the overall process, frequencies which need bear no relation whatsoever to any environmental periodicity. Thus, for example, the cell generation times of bacteria, being of the order of 20 min to a few hours depending upon the nutrient supply, bear no known relation to any environmental periodicity such as the day/ night cycle. Also, the menstrual cycle is not clocked in to any environmental cycle, despite romantic conceptions about the moon. However, there are many physiological and behavioural rhythms in organisms which bear a direct relation to environmental rhythms, and appear to have originated as a result of adaptation to a periodic environment. Such rhythms have an extrinsic rather than an intrinsic origin. The most intensively studied of these are the 24-h rhythms relating to the solar day and 12-h tidal rhythms in marine (usually littoral) organisms.

The literature on the behaviour of these biological clocks is now very extensive (Cold Spring Harbor Symposium **25**, 1960; Aschoff, 1965; Bünning, 1964). A central problem here is to establish whether or not any observed behavioural rhythm is generated by the organism itself and gets locked or entrained to the environmental periodicity at a particular phase or whether it is the external rhythm which causes the periodicity in the organism directly. There is still considerable debate about these alternatives, but it has been conclusively established that for the case of some circadian rhythms at least the periodicity is generated within the organism, independently of any periodicity in the environment (De Coursey, 1960). However, it is not usually disputed that the evolutionary origin of the behavioural rhythm in such cases is to be found in the adaptive response of the organism to an environmental periodicity. Innate behavioural rhythms are then regarded as conferring a greater adaptive advantage on the organism, which can anticipate environmental periodicities even if for one reason or another it is temporarily isolated from the normal environmental cycle.

If the frequency of an intrinsic rhythm, such as that of the cell division cycle, happens to fall within the range of an environmental periodicity such as the solar day, then we may expect that there will be some kind of interaction between the rhythms, resulting in a quite complicated periodic organization which is to be understood as partly adaptive and partly of internal origin. This in fact often appears to have happened with cell division times, particularly for the different cell types of multicellular organisms. Bruce (1965) has demonstrated a rather striking convergence of all generation times in the neighbourhood of 24 h, or multiples thereof. The organism having itself adapted to a basic 24-h periodicity with, for example, various circadian hormonal cycles, it appears that the growing and dividing cells in different tissues such as skin, smooth muscle, gastric mucosa, etc., have also become adapted to 24-h periodicities in the *milieu intérieur*, their environment. Thus we have intrinsic rhythmicity adapting to and interacting with rhythms of extrinsic origin, with attendant complications for the analyst.

Interactions of this kind may occur not only when frequencies fall within a similar range, but also when there are simple multiple relations between frequencies. Thus 24-h cycles can interact with a 28-day menstrual cycle and vice versa so as to give maximally coherent total behaviour. Ovulation normally occurs at a particular time of day; while cell division rates in the ovary vary with the 28-day cycle. This nesting of frequencies may be expected to occur throughout the frequency spectrum which can be very extensive, ranging from monthly and seasonal rhythms to the high frequencies of neural control processes such as eye tracking movements (Young and Stark, 1965). A description of this kind of mutual adaptive ordering of commensurate frequencies has been given by Iberall (1969).

We have here an example of a fundamental aspect of biological organization, the hierarchical ordering of processes at different levels which are related to one another functionally by virtue of specific control relationships. The concept of functional hierarchy discussed by Pattee (1970) in a very interesting analysis of the nature of the biological process is, we believe, exceedingly useful in the investigation of the way in which one level of activity is functionally embedded in another. This topic is beyond the scope of this introduction. However, it is clear that one cannot venture into any biological domain, including biological rhythms, without encountering the intricate interplay of time-scales which is so characteristic of biological behaviour, and can be such a source of confusion to the conscientious experimenter. In the study of rhythmic phenomena, the interaction of different frequency domains can cause quite severe analytical difficulties, especially if the oscillations are non-linear. A discussion of some of the problems presented by hierarchical temporal organization and non-linear oscillations in biological rhythms has been presented by Goodwin (1963).

## B. *The dynamics of biological stability*

The arguments presented above suggest the logical origins of biological rhythms, but not their dynamic nature. Oscillations which depend directly on an environmental periodicity, such as the diurnal rhythms of photosynthetic activity in plants, are very easily understood as the direct response of the system to an extrinsic cycle. However, innate rhythms such as the activity rhythm of the mouse (De Coursey, 1960) or the cycle of cell growth and division of any cell type requires an explanation in terms of some internal, autonomous oscillator. In no case has the dynamic nature of such an oscillator been established, but it is generally assumed that the oscillation arises as a consequence of the operation of a control circuit of some kind operating on the principle of negative feedback. Such control devices are the basic stuff of physiology, and it is common knowledge that negative feedback control can readily result in oscillations in system variables (see Section IV C). Thus it is natural to postulate that the two phenomena are intimately connected.

Some direct justification for this postulate follows from the work of Chance, Hess and Betz (1964), and Betz and Chance (1965) on a biological oscillator which has been extensively studied recently and which does depend in an essential manner on negative feedback. This is the glycolytic oscillator of yeast, which has been reconstructed in the test-tube and investigated directly. It consists of the enzymes involved in the degradation of trehalose, a disaccharide characteristic of yeast, to ethanol and $CO_2$. When these enzymes are mixed in a test-tube in their proper relative proportions together with a supply of trehalose, an oscillation in the concentrations of the metabolic intermediates occurs which continues until the trehalose has been exhausted. This oscillation has been shown to depend crucially upon an enzyme called phosphofructokinase, a key controlling enzyme in the sequence which is sensitive to feedback inhibition by adenosine triphosphate, an end product of the glycolytic pathway. The frequency of these oscillations is of the order of one per minute. Thus the direct connection between feedback control and oscillatory activity in a biological process has been demonstrated and studied in this case, and there is no difficulty in extending these principles to quite a wide class of feedback control systems. Different frequencies arise from systems in which the time constants for the constituent steps are either smaller or greater than those for the enzyme-catalysed steps involved in glycolysis. Very long-period oscillations such as those occurring in the menstrual cycle evidently involve the relatively slow processes of cell maturation and tissue growth, while very short-term oscillations such as those in the central nervous system of the embryo (Hamburger and Balaban, 1963) or the adult (Hoagland, 1936) involve primarily ion fluxes across membranes.

It would undoubtedly be a mistake to insist that wherever an oscillation is observed there must be a well-defined, isolable control pathway with negative feedback which is the basic generator of the oscillation. The dynamic behaviour of biological systems may arise from properties of a complex network of interacting elements rather than collections of single control units with weak coupling between them (Kauffman, 1969). Given such a network, it is often very difficult and may be quite impossible to give an exhaustive and unique dissection into units of control out of which the whole system can be reconstructed. One of several techniques (see Section V C) that can be employed in attempting to unravel the degree of autonomy of various observed oscillations or rhythms is to perturb the system in various ways from some defined steady state of behaviour. For example, one can disturb the habitual 24-h sleep–wake cycle in a diurnal or a nocturnal animal by imposing a new frequency such as an 18-h day and observe how rapidly different physiological variables become entrained. Such experiments with humans show, for example, that the rhythm of body temperature adapts very quickly to the new régime, whereas the rhythm of renal excretion is slow to change (Lobban, 1965). This observation tells us that the control circuits within which we may consider these variables to be embedded are distinct one from the other, a control circuit now being defined as a set of variables which are strongly coupled to each other and change in a correlated manner such that some quantity remains in a defined neighbourhood of a fixed point or bounded trajectory.

One may then perturb the system in the opposite direction, putting it back into the normal 24-h periodicity and seeing how rapidly certain variables return to normal. Such perturbations cause dynamical dissociations in the system, revealing the points at which it is most easily fractured, thus uncovering the partially autonomous dynamic sub-systems which together constitute the whole. Such decompositions of the system are not necessarily unique. Different perturbations may reveal slightly different units. However, the procedure does often allow one to reduce the whole to a set of interacting subunits out of which the total systems can be reconstructed, thus providing the analytical basis for model-building.

Having identified and to some extent isolated a particular rhythm from its dynamic substratum, it is important for the purposes of model-building to identify its stability characteristics. Different types of oscillator have different types of stability. The relaxation oscillation is one which has a fixed amplitude of variation and variable frequency: it gives a one-variable clock. A limit cycle oscillator has variable amplitude and frequency, but for fixed conditions it always returns to a fixed trajectory, the terminal limit cycle, after a perturbation. A conservative oscillator such as the classical pendulum showing simple harmonic motion has variable amplitude but a fixed frequency, so it may be regarded as giving a stable frequency even after small pertubation—and so on. Quite general topological-type or phase-plane representations of these different types of oscillator and their stability have been given in books such as Minorsky's "Non-linear Oscillations" (1962). An elegant example of the use of these general, basic dynamical ideas for the design of experiments to identify the nature of a particular oscillation has been given by Winfree (1970) in his analysis of the rhythm of emergence of the fruit fly, *Drosophila*, from the pupa, known as the eclosion rhythm. The analysis in this case was complicated by the fact that the rhythm observed occurs at the population level, not at the level of the individual, while it is the nature of the clock which is functioning within the individual organism that is being investigated. However, by observing the responses of a population of pupae to specific stimuli, which in this study were flashes of light delivered at particular times after the initiation of the population rhythm, Winfree was able to characterize the stability properties of the rhythm or clock in the individual organism.

The next step in the reconstruction of a rhythmically-organized system, producing a model for the interactions of oscillating sub-systems, is a very difficult one. As nearly all biological oscillations are non-linear, rendering Fourier or other linear analyses inapplicable because the superposition principle is violated, an alternative approach is required for both the decomposition and the reconstruction of the system. The theoretical study of interacting non-linear oscillators is still at a relatively early stage of development. The main emphasis here is on defining the qualitative features of this behaviour, which are revealed as stability properties relative to phase, amplitude, and frequency. This emphasis on stability is appropriate, since it is the overall stable organization of the living system, within which development, adaptation, and learning take place, that we are attempting to define when we study its dynamical behaviour. We have seen that biological rhythms appear to be basic to this dynamical organization of the total system, allowing for the stable temporal resolution of simultaneously incompatible processes within the system and the optimal adaptation of the system to periodicities in the environment. But it may be some time before we have a fully-developed theoretical framework within which this behaviour can be adequately represented and analysed.

## IV. Functions of Rhythms

In the previous section a general introduction to the occurrence of rhythmicity in biological systems was given in order to stress its pervasiveness, its importance and its general properties. In this section some examples of how these properties of rhythms can influence behaviour, and particularly human performance, will be discussed in relation to their roles in timing, prediction, and control.

### A. *Timing*

A possibly important function of rhythms, related to human performance, is that of timing and synchronization of interrelated events. The brain is a sophisticated computing device, containing some $5 \times 10^{10}$ neurones. If it operates in any sense whatever like the computers with which we are familiar, it must have some means for ensuring that sequences of events occur in the right order, and that related processes interlock temporally. These require timing, and timing is difficult to imagine without some oscillatory process.

In most digital computers there is such a synchronous process, or clock which ensures that events are initiated in the interstices between storing or retrieving data or performing logical operations. This prevents anything new from happening until a previous operation is finished. One hypothesis about brain function which has relevance to performance is the proposition that the brain works in just such a synchronous fashion. There is perhaps some minimum time for an event, a decision, a percept or a motor output, and everything happens in integral numbers of these quantal events, or phsychological moments. Stroud (1955) has been instrumental in popularizing this idea, though similar notions had been in existence for many years. A full blown theory of psychological moments holds that all brain activity is organized into rather discrete events slotted into quantal divisions of time, a moment representing not only the psychological equivalent of a computer's core cycle time, but also, as Efron (1967) has put it, "the duration of the present".

Rather more popular than the all-embracing psychological moment have been more limited and circumspect versions, notably that of the *perceptual* moment, in which intake of stimuli is sampled. Perception according to this hypothesis includes a process like a cinema film, running at (say) 10 frames/sec. Within any frame only the average of any parameter that varies during that time can be recorded, and only at the end of the period can data collected during that period be passed on for further analysis. White (1963), amongst others, has espoused a relatively high frequency (10 Hz) version of this theory, while Broadbent (1958) at one time, but no longer (Broadbent and Gregory, 1967) preferred a 3 Hz version.

The obvious attraction of this type of hypothesis is that it would make the temporal organization of perception, and its integration with related events much easier to understand. In particular the problem of deciding when some new perceptual event has occurred requires minimal computing effort in a sampled data system (Shallice, 1964). Despite the attractions of the hypothesis, however, the evidence for both the psychological moment and the perceptual moment is not as strong as it might be (see Chapter 5).

Despite the fact that empirical support for untriggered periodic moments is not unanimous, the concept of timing of related events remains of importance. Even if there is no inflexible internal clock which times the length and sequence of perceptual processes, there are still complicated sequences of logical operations occurring in perception and memorization, and these need to happen at proper times in relation to each other. The steps in perception and short term memory discussed for instance by Neisser (1967) all require some mutual synchronization of events. Although these may not be timed on a basis of an exactly repeating periodicity, many of the processes have time-brackets associated with them; time to fill a store, scanning time, time to switch attention and so forth. Some of these processes may be triggered by stimulus events rather than by a clock; furthermore, the time associated with a process may be the time for it to reach an end point, rather than for a clock to run out. This for instance is the case in Welford's (1952) account of the psychological refractory period. The synchronization of many psychological events may turn out to be of this kind; related events being held up until some operation is finished, rather than until some ancillary oscillator completes its cycle. Nevertheless there may be a place for clock-timed processes, triggered by an incoming signal and then lasting for a more or less predetermined period.

For many purposes this difference may not be important. In either case limits will be set upon the speed of performance because input of data, the making of decisions, or the emission of responses is intermittent. The essence of the problem put by the single channel hypothesis of human performance, and the effects of attention on performance is that in these systems of mixed parallel and sequential organization of information processing, alternation or oscillation between different states is the rule. Even though the oscillations may be irregular, many of the principles of periodic processes hold, and there may well be a more or less rhythmic variation in the times at which things can and cannot occur.

An important theoretical point is that for events between which a fixed time relationship is required, a system of weakly interacting non-linear oscillators is ideal (Winfree, 1970). Such oscillators will readily entrain upon one another, at the same frequency, as well as at higher or lower harmonics as suggested in Section III. The uses of such devices in stabilizing temporal relationships are clear, and Van der Pol (1940) has pointed out the significance of relaxation oscillators in biology in this context.

The sort of thing involved in stabilizing the phase relationships of rhythms which are harmonics (or sub-harmonics) of one another can be appreciated as follows. If one beats rhythmically with one's right hand at almost any speed it is easy to beat with the left hand at sub-harmonics of this frequency, e.g. every second right hand beat, every third, or whatever. What is not so easy is to make 3 regular beats with one hand occupy the same time as 4 regular beats with the other. Maybe this betrays something of the organizational structure of motor-control processes. Certainly at least since Lashley's (1951) famous paper on serial order in behaviour, the importance for the control of movement of endogenous timing and sequencing (which must in some sense be built up on internal clocks) has been clear.

Furthermore, there is physiological evidence that the brain has intrinsically oscillatory arrangements. For instance, low frequency stimulation of the brain has for long been known to produce sleep (Akert, Koella and Hess, 1952), and externally applied rhythmic stimulation is the principle upon which is based the Electrosleep machine, an important piece of equipment for inducing sleep and treating mental illness in the Soviet Union. These methods probably operate by entraining internal oscillatory circuits which encourage sleep. Indeed, the periodic nature of EEG patterns, particularly during sleep and inactivity, indicates the presence of neuronal oscillations. Photic driving of EEG rhythms and the relation of these to performance is discussed in Chapter 5. Neurological entrainment of a somewhat more sinister kind occurs in the precipitation of epileptic seizures by flashing lights at the right frequency. It may also be that the rhythmic properties of music are pleasant because of their potentialities for entraining brain rhythms. Besides indicating the existence of entrainment in the brain, these phenomena, particularly the possibility of stimulating sleep by rhythmic stimulation, need to be borne in mind in applied problems of human performance. Since many work situations involve revolving or reciprocating machinery emitting both sounds and vibrations at frequencies quite suitable to encourage sleep, thought might perhaps be given to abolition or desynchronization of the effective frequencies in work settings.

One might suppose with Pringle (1951) that many neural processes are built up from non-linear oscillations. Certainly the very regular periodic firing of action potentials in axons is a fine example of a relaxation oscillation, and there is no doubt that the EEG has rhythmic features. We might then regard brain physiology as consisting of the interactions of coupled oscillators, some sensitive to external events, others endogenously active. Meaningful computation in such networks has been shown to be possible because of the co-operative phenomena which result from the entrainment and triggering properties of the constituent oscillators (Cowan, 1969). Thus not only from the point of view of psychology, but also from the point of view of neurophysiology comes the possibility that fundamental insights into brain mechanisms will depend upon the understanding of rhythmicity. Whether rhythms will turn out to be at the bottom of problems like the minimization of computing effort in the brain, by means of periodic sampled data schemes, or whether problems such as the multiplexing of incompatible activities are solved by entrainment of weakly coupled oscillators, is not known. It probably is the case that this kind of theoretical approach has considerable possibilities. It may also be the case that the occasionally confused and unsatisfactory state of research into periodic processes in the brain results from lack of understanding of the properties of oscillators; for instance the way in which they can respond to periodic stimuli of low amplitude, but are blocked by larger stimuli.

## B. *Prediction of periodic events in the environment*

A slightly different kind of problem occurs when an organism needs to predict the state of the external environment. As Craik (1943) has pointed out, this entails the brain being a model of the environment, and one way in which this is achieved is through learning. There are other methods of prediction which do not require learning however. For periodic events in the environment models would consist of endogenous oscillators. The chief instances of this are the diurnal, or circadian rhythms which depend upon oscillators modelling the revolution of the earth, and the sequence of night and day.

(Terminology for 24-h rhythms is used differently by different authors. For some circadian and diurnal are synonymous terms. Others distinguish between circadian meaning having a period of about 24 h, and diurnal with the additional implication of being synchronized by the lighting sequence. Whatever term is used, these rhythms should be distinguished from mere light-dependent processes, since they depend upon internal clocks synchronized to the light–dark cycle. When the synchronizing signals are removed the clocks or oscillators continue more or less accurately on their own.)

Diurnal rhythms show other interesting properties of the kind associated with non-linear oscillators. In a single individual several diurnal processes may each be driven by a separate oscillator, and in the normal course of events they are all entrained with one another. They can, however, be separated by maintaining one of the processes at a constant level, when the other rhythms will continue uninterrupted. Furthermore, phase can be reversed, by reversing the normal light and dark periods. For most diurnal rhythms this takes a week or so. Internal oscillators can also be synchronized on dark–light periods different from the normal 24 h. Artificial day lengths from 16 h to 32 h can be accommodated by many circadian oscillators.

Very few animals operate equally well by day and night. For most their ecological niche includes being active either at night or during the day. At the simplest level internal clocks serve to wake the animal up in preparation for its period of activity. The onset of dawn or dusk would not be a sufficient signal for this, because the animal may not be in a position to receive the signal. However, if the light, or dark signal is used to synchronize an internal oscillator, then it does not matter whether the animal sleeps in a dark hole, well away from daylight, or if it is unusually gloomy on some mornings. The internal clock does the main business of timing, with environmental light and darkness simply keeping it in synchrony with the rhythm of terrestrial rotation.

This daily cyclicity of sleeping and waking, as well as other associated behavioural and metabolic rhythms (see Mills, 1966, for a review of human physiological rhythms) is obviously of importance both in theoretical and applied problems of human performance. There is no particularly convincing theory of why man spends a third of his life asleep. The idea that sleep is used to regenerate metabolic substances used up during the day, though it has not received firm experimental support, is still under consideration (Webb, 1968, and in Chapter 4). Perhaps the most likely substance to be used up during the day, and stored during the night is ATP, which is essential to energy release in most mammalian tissues. Webb (1968) discusses the evidence that the availability of ATP declines as the day continues and that it is regenerated most efficiently during sleep. An alternative theory is that sleep is necessary because the brain, rather like a complex computer, periodically has to be taken off-line in order to sort through resident programs, update them and dispose of obsolete material which would otherwise occupy valuable storage space and promote inefficiency (Newman and Evans, 1965). Maybe the reason why sleep is necessary is simpler. Perhaps man sleeps simply because he has evolved from animals which can only operate successfully during the daytime. Sleep is certainly a good solution to the dual problems of keeping relatively still in a place of safety, and avoiding the boredom of prolonged inactivity (but see Chapter 4 for a more informed opinion).

Whatever the purpose of sleep, its existence and the existence of associated rhythmicities is undoubted. The difficulties produced by a reversal of the light–dark sequence or of the sleep–waking routine are known to anyone who has flown the Atlantic or who has worked on a night shift, and these issues are discussed in Section V C, and in Chapter 2.

One obvious problem in studying human performance is that of recognizing rhythmicities in order to be able to co-operate with them. Systematic variation of efficiency during waking has been recognized (see Chapter 2). Optimal performance may occur at different times for different people (see Chapter 3) and this fact clearly should be used, rather than ignored. The issue also exists in purely experimental work on human performance as well as in applied situations. Presumably if a given experimental subject were always tested at the same time of day the variance of experimental results would be diminished.

Changes in performance may occur during the day because the oscillator controlling sleep–wakefulness has an output which is not a square wave. Since arousal is an important construct in many theories of performance, the oscillatory nature of arousal may be essential to understanding what goes on, rather than an annoying source of variation. One might expect, in common with diurnal rhythms of other kinds to see an endogenous systematic trend during the day, with at least one peak at some point. This is exactly what empirical data do indicate (see Chapter 2). If one continues to suppose that it is arousal that accounts for variation of performance as a function of time of day, one might also expect to find a higher frequency variation superimposed upon the 24-h cyclicity. In Webb's (1968) view paradoxical, or rapid eye movement (REM) sleep is a device to substitute for waking activity in the brain, and to avoid changes of thresholds and other neural parameters which would occur on functional withdrawal of stimulation. The paradoxical phase is well known to occur periodically about every 90 min during sleep, as illustrated in Chapter 4, Fig. 2.

The easiest way in which such a pattern might be built up is with an internal oscillator which generates REM with a period of about 90 min. One may suppose that this oscillator runs all the time, but that its effect on the brain is gated on and off by the slower 24 h oscillator. This raises the interesting question of whether the gating is complete or whether the 90 min periodicity of REM periods in sleep can be detected during the daytime. Othmer, Hayden and Segelbaum (1969) report that periods of REM and diminution of muscle tone do occur in subjects with their eyes closed during the daytime. It remains to be seen whether this 90 min rhythm exists during active behaviour, but if it is an integral part of neuronal organization, it well might affect arousal and performance (this issue is further discussed in Chapter 2).

In summary, since diurnal rhythms have a function which is at least that of predicting daytime, and since the nature of the oscillations is such that the waveform is not simply on–off, then there is good reason to expect slowly varying arousal during the day, and hence corresponding changes in performance.

Prediction of daylight is only one possibility of rhythms. Predictions of tides and seasons are well known in the animal kingdom. Perhaps more interesting from the point of view of human performance are oscillations tuned to higher frequency patterns. If an animal is habituated to a regular series of stimuli, measures such as the orienting response, and single unit responses wane. If one of the regular series of stimuli is omitted, there is a response, as it were, to the absence of the stimulus. This is only easily explained if some internal oscillator is set in motion by the repetitive stimuli, whether the oscillator is identified with Sokolov's (1960) neuronal model, or not (Horn, 1967). It also implies that neural oscillations can be set up at any frequency over quite a wide range. There seems no reason to doubt that prediction of regularly repeating patterns, and detection of deviations is an important perceptual ability. Much the same thing is true in the spatial domain. Gestalt psychologists were interested in regularities and irregularities of repetitive patterns and these also require some type of internal iterative process in order to predict, and hence detect irregularity.

## C. *Oscillations as part of control processes*

Many simple controlled processes are thought of as being regulated to a steady state. It is of course well known that for any feedback controller the response is subject to oscillations because of delays around the feedback loop. If the input to a simple negative feedback controller is rhythmically varied with a period that is equal to twice the delay around the feedback loop, then the system is set into vibration. Linear controllers are specifically designed so that the gain, or amplification around the feedback loop, at this frequency is less than one. If it were not, then any noise in the

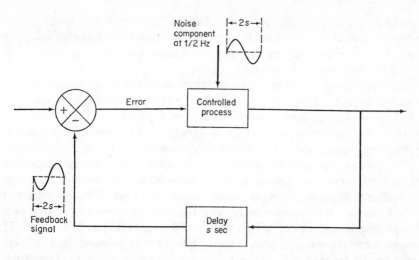

FIG. 2. A negative feedback control system, illustrating the way in which oscillations can develop in it. When delayed by *s* sec the noise component, with a period of 2*s* sec, adds to itself rather than being subtracted.

system which perturbed the controller would start it oscillating at its own natural frequency (see Fig. 2). In this type of controller the output is fed back to the input, and subtracted from it (negative feedback). If any oscillation with a period of twice the feedback delay occurs this will mean that an oscillation reaches the input from the feedback loop exactly out of phase with the input. A sinusoidal oscillation phase shifted by 180° is exactly the negative of the input, but since the device subtracts this feedback signal, the effect is to subtract a negative signal, and the result is addition. The oscillation at this frequency therefore adds to itself, since with a loop gain of more than one, the feedback signal is larger than the input. It grows exponentially until either the device is wrecked or saturation occurs. This situation in a linear controller is known to the engineer as instability, and a principal object in the design of most such controllers is a rapid rate of response, and hence high loop gain but without instability.

Biological systems, insofar as they behave in any sense like linear servo systems, have been designed, presumably by evolutionary processes, to be very stable, i.e. with loop gains very much less than one at the critical frequency. One can occasionally make them go unstable and oscillate by increasing the loop gain. This has, for instance, been done by Stark (1959) in his work on the human pupil servo-mechanism. If light in the form of a thin beam is shone on the edge of the pupil, rather than diffusely across the whole eye, then any change in pupil diameter will have a much larger effect on retinal illumination than in the diffuse case. This is equivalent to increasing the loop gain of the system, and the result is a vigorous oscillation of the pupil at its natural frequency.

Inherent noise in the system can be shown by Fourier analysis to be equivalent to a mixture of sinusoidal oscillations at all frequencies. One of these will contribute a small oscillation at the critical frequency for oscillation. With each passage of this signal round the loop it will grow if the gain is greater than one, and the oscillation will increase.

We do not normally see oscillations which are evidence of instability in biological systems, although certain pathological conditions involving disturbance of motor control are characterized by oscillations or tremor which may arise from malfunctioning feedback systems. Oscillation at a lower amplitude can occur in any servo-like control system as a sort of resonance if the input at the critical frequency is large enough. The 9 Hz peak in the frequency spectrum of human muscular tremor (Hammond, Merton and Sutton, 1956) may be an example of this. It has the theoretical importance of being possibly able to provide evidence about the control systems involved in performance of manual skills. For instance, Sutton and Sykes (1967) recorded the error power spectrum (see Fig. 6) from performance in a tracking task, where subjects had to maintain a spot centrally on an oscilloscope screen by pressing against a loaded joystick. The 9 Hz tremor disappeared when the subject closed his eyes, and Merton, Morton and Rashbass (1967) extended this finding by showing that the frequency of tremor shifted when extra delays were incorporated into the visual feedback loop. These data clearly have something to do with the control loop involved in visual monitoring of the task, and Eccles (1969) has recently treated these results as evidence for his "dynamic loop" hypothesis of the role of the cerebellum in motor control. This illustrates the important principle that the presence of control circuits can be betrayed by the oscillations they produce, and the oscillations can be used to investigate the nature of the control process.

Another possibility has been put forward as an explanation for the oscillatory behaviour of the operator's output. Young and Stark (1965), Wilde and Westcott (1962), Venables (1960), Latour (1967) and a number of others have assumed a periodic sampling process of a kind mentioned earlier in the context of timing, and discussed in Chapter 5.

The tracking task has been the main example of human performance in which frankly rhythmic behaviour has been explored in relation to a control process. The hypothesis of central intermittency arose from Craik's (1947) observation of an approximately 2 Hz periodicity in continuous tracking waveforms. Several models of human tracking behaviour have been put forward and these have been reviewed by Licklider (1960), Young and Stark (1965) and Fogel (1963).

The concepts mentioned so far, however, barely scratch at the surface of possible roles of oscillatory processes in control systems. All of the systems discussed above are based on quasi-linear servos, in which oscillation has been attributed either to an inessential consequence of the mode of construction of the device, or to periodic sampling. A much more important possible role for oscillatory processes in biological control is in non-linear controllers which have no reference levels or set points. A substantial problem in understanding biological regulators is what constitutes a set point. Body water, for instance, is rather closely controlled, and one can make servo models of the control process with drinking taking place when the actual value of body water subtracted from some reference value exceeds some threshold quantity (Oatley, 1967). It is, however, not easy to imagine what would constitute a reference level for this process of body water regulation. It may be that oscillations between two loosely defined points in non-linear controllers with delays in the loop may play the part in biology that is played by set points in servo regulators. It is quite possible that the problem of reference levels is solved in some biological controllers in this way, in which case understanding the variations of controlled parameters would be essential to understanding how the control operated.

Another possibility that has barely been explored is the use of oscillations to determine the operating point of the system. Some sophisticated artificial controllers make use of small signals of a defined type, either white noise or some oscillation, which are deliberately injected at the input of the controller. The output variations are then correlated with this input perturbation signal. The procedure can determine an aspect of the transmission characteristics of the controller, and the result is used to alter the transmission characteristics towards some optimal state. Often some performance index is derived, such as a cost function which must be minimized. An example of this would be a system in which efficiency was an inverted U-shaped function of some parameter. By injecting a small perturbation signal, it would be possible to detect which side of the inverted U, or hill, the process was operating upon. This might be a simple matter of finding the first derivative, or slope. A hill-climbing scheme of this kind may be used in the control of accommodation in the human eye. Any degree of de-focus of the lens produces blur, and Alpern (1958) has suggested that in order to disambiguate the blur, and find out which direction of lens movement improves the focus, the accommodative mechanism undergoes a continual oscillation. The lens does in fact show such a maintained oscillation in lens power, even when viewing a stationary target.

Although accommodation can be strongly driven by convergence cues, it can take place with only image blur as the stimulus, and in this situation the perturbation method may be used to find the operating point of the system, and thus to control it in the right direction.

Control processes of this kind have been postulated in other contexts. Priban and Fincham (1965), for instance, argue that movements of the respiratory muscles are controlled using the rhythmic variation of $pCO_2$ in the blood brought about by periodic breathing. This oscillatory signal defines the position of the system on a hill relating efficiency of gaseous exchange to the energy expended in making breathing movements.

This kind of adaptive principle has not been widely applied to biological control systems, partly because the theory of adaptive control is not well understood (at least by biologists). Simple linear servo theory for biological systems will probably be replaced by theories of self-adaptive control which may include oscillations as important elements of the process. It is clear at least that many control processes, especially in situations such as skilled performance, are self-adaptive. Learning a skill with use of knowledge of results is similar to injecting trial and error pertubations into the system, and then modifying performance parameters in a direction that improves the behaviour. Engineers as well as psychologists would like to know how this is achieved, since it is now thought that some complex industrial processes can only be successfully controlled by self-adaptive (learning) processes of this kind.

Even in highly overlearned skills the perturbation method of finding the present characteristics of the system may be important. Racing drivers, it is said, waggle the steering wheel on approaching a corner so that they can adapt their behaviour to the road conditions at that particular place and time. We may expect to see therefore increased attention devoted to adaptive control theory, particularly in the analysis of skilled human behaviour. This may result in many of the small rhythmicities and oscillations which are difficult to understand at present, being recognized as the perturbation signals used for determining the state of the system.

## V. Special Methods for Investigating Rhythms

### A. *Signal analysis and the identification of rhythms*

Quite often an experimenter wants to know simply whether a signal contains an oscillation, that is to say, whether there is any significant periodic component in that signal. A related problem is whether a system tends to transmit a certain frequency at a greater amplitude than that of neighbouring frequencies, i.e. are there phenomena of a resonant kind in the system? These problems will be treated separately.

1. *Fourier analysis.* It is well known that signals can be represented as the sum of sinusoidal oscillations of specified amplitude and phase. Thus a brief impulse can be represented as the sum of sinusoids of all frequencies, at equal amplitudes, but of a phase such that they all add at one point, and cancel out elsewhere. More complicated signals than impulses can also be represented as the sum of the sinusoidal frequencies that they contain, and most signals which can be thought of as either repetitive or lasting for a finite time can in theory be represented as functions of frequency. The phase and amplitude of each component plotted as a function of frequency is known as the spectrum.

There is a formula for converting a signal from a function of time to a function of frequency. This is called the Fourier transform, and it is usually written

$$G(j\omega) = \int_{-\infty}^{\infty} f(t) . e^{-j\omega t} \, dt \tag{1}$$

This is simply a mathematical recipe for finding the equivalent function of frequency, $G(j\omega)$ from a time-varying signal $f(t)$. Both representations exactly and uniquely specify the signal. For many purposes it is more convenient to know what frequencies a signal contains than to know its variation with time.

Formula (1) can be interpreted as follows: it says simply that one can find the frequencies in a signal by taking the signal $f(t)$ and multiplying it by sinusoids $(e^{-j\omega t})$ at all frequencies $(\omega)$ and integrating the result over all time. In a simplified version we might take the signal as in Fig. 3, and sample it as shown. At each of these regular sampling intervals we multiply the value of the signal by the value of the sinusoid, and continue to do that throughout the length of the signal. We then integrate, that is to say in this sampled version add, all of these products, and the result gives the amplitude of the frequency spectrum at the frequency of the sinusoid used. This can then be plotted as a point in the frequency function. The reason why this multiplication and integration process extracts a frequency component from the signal is that if any 2 pure sinusoids are multiplied and integrated (added) in this way, the result will always be zero because the positive and negative products on average exactly cancel out, except when the signals are at precisely the same frequency, when a non-zero result will be found. Thus if the signal $f(t)$ has the equivalent of a sinusoid at what might be called the probing frequency, then there will be a non-zero product proportional to the size of only that particular frequency component.

In order to obtain the whole spectrum one obtains sums of cross products for a range of frequencies $(\omega)$ at intervals from low to high frequencies.

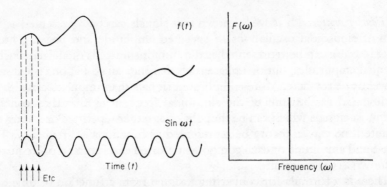

FIG. 3. Diagram to illustrate the process of calculating the Fourier integral. On the left, a signal $f(t)$ for which the spectrum is to be found, and a sinusoidal signal sin $\omega t$, are both sampled regularly at sample points shown by the arrows and dotted lines. The signal $f(t)$ and the sinewave are now multiplied and summed as specified in formula (1) (see text). The result, plotted as the height of a line on the spectrum at the frequency $\omega$ of the sinusoid used, is shown on the right of the figure.

This basically is what formula (1) says, except that the integral sign represents a continuous rather than a sampled process. More practically it should be replaced by the summation sign $\Sigma$, and when performing the calculation on a computer this in fact represents what one actually does.

The only other point about the formula (1) is that the expression

$$e^{-j\omega t} \tag{2}$$

represents a sinusoid specified in both amplitude and phase. In order to specify completely a sinusoid not only its frequency, but also the phase, or point in its cycle that has been reached at any given time must be known. Conventionally this is done using a complex number as in expression (2). If ordinary positive and negative numbers are a means of representing points in one dimension, i.e. along a straight line, with zero as an arbitrary origin, then complex numbers are a means for representing a point in 2 dimensions, or on a plane. A 2-dimensional number is therefore what is required to represent 2 orthogonal parameters—phase and frequency of a sinusoid—and this is what the expression (2) does (there is no space for further elaboration, but for a simple though very good introduction both to complex numbers and to calculus of the kind used in signal analysis see Ritow, 1963).

This can all be done relatively easily in a digital computer. The spectrum then obtained will show any prevalent rhythmicities as humps in the amplitude spectrum. For instance, the spectrum of male speech displays prominent frequencies around 500 Hz, but with a small, but important group around 6000 Hz.

It is of course possible to calculate the Fourier transform of signals without a computer. For signals that can be recorded electrically a normal way is to play the total signal through a bank of sharply tuned filters. This method is widely used in EEG analysis and in making speech spectrograms. Each resonant filter does the equivalent of multiplying the signal by that frequency as in formula (1), and the amplitude output from each can be plotted as a point on a frequency spectrum. Some caution is necessary, however, in any frequency analysis technique. The theory of Fourier transforms defines sinusoids as extending infinitely in the time from $t = -\infty$ to $t = +\infty$. This is impossible to realize in physical systems. When computing one can deal with the problem by making sure that the signal is zero at either end of the run, so that the cross products are zero when one stops calculating. With a continuous signal and filter analysis one must make long enough runs to ensure that the transient warm-up effect of switching on the signal has died away before measuring the amplitude of the output. With the filters that are used for this kind of task this takes at least 10 cycles at the filter frequency. One cannot measure the instantaneous frequency of a signal or get an accurate statistical estimate of its frequency content by looking at it over a period which is shorter than several cycles of the frequencies of interest.

2. *Autocorrelation.* There is an alternative method for finding the frequency content of signals if one is not concerned about phase information, as for instance when one wants to know simply whether a rhythm exists, and what its frequency is. Once again access to a digital computer is desirable. One calculates first the autocorrelation function, and secondly, though one need not necessarily do this, the power spectrum.

The autocorrelation is formed by a product of the signal with a delayed version of itself. The formula is

$$r(\tau) = \int_0^\infty f(t)f(t+\tau)\, dt \tag{3}$$

In the computer one proceeds much as for the Fourier transform. Summation of regularly sampled points is used instead of integration. Thus one samples the signal at some suitable frequency; in order to lose no information the sampling theorem states that this frequency needs to be no lower than twice the highest frequency contained in the signal. Since one may not know exactly what this is, one has either to make a guess, or to decide what frequencies one will be interested in, and then sample at twice the rate of the highest one. Products between the sample points and the same points on a second representation of the signal are then calculated at a delay $\tau = 0$. One then shifts (as it were) one version of the signal along by one sample interval, and again sums products of the signal with corresponding points of this delayed version. Figure 4 shows this process diagrammatically. It is important to have a sufficient quantity of the signal to avoid "end" effects, and also either to multiply the same number of samples together for each value of $\tau$, necessitating that the delayed version be at least twice as long as the stationary version, or else to use a weighting factor inversely proportional to the number of products at each delay.

As an example, Fig. 5 shows autocorrelations of the small vibrations of the human pupil as found by Stanten and Stark (1966). This function is not untypical of a biological signal that fluctuates somewhat irregularly. The peak at $\tau = 0$, which is conventionally normalized by assigning to it an amplitude of one, is the highest point in any autocorrelation function. This represents a value for the square of the signal. For anything less than a perfectly repetitive signal this function must be less than one at all other values of $\tau$. (This is obvious when one thinks what the autocorrelation means: it is a function representing how much the signal has in common with various later parts of itself. Clearly it has most in common with the part nearest to itself, i.e. when the delay $\tau = 0$.) For a noisy, or irregular signal the central peak tends to be tall relative to the rest of the function. The more random the signal the less it is possible to predict what it is going to do next from any given point and the narrower the central peak

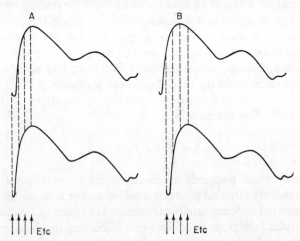

FIG. 4. Diagram to illustrate the process of autocorrelation. At A the signal is sampled as indicated by the arrows, and products taken and summed (see text). As shown at B the next step is to delay one version of the signal relative to the other by one sample interval and to repeat the process.

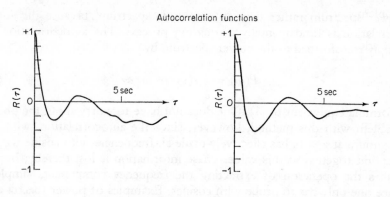

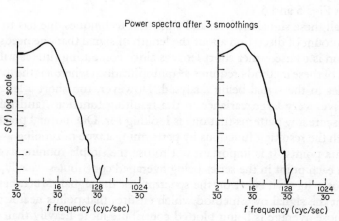

FIG. 5. Autocorrelation functions and the power spectra derived from them for the oscillations of the human pupil under 2 conditions of illumination. Reproduced with permission from Stanten and Stark (1966).

of the autocorrelation function is. If there is any periodicity in the signal, then this manifests itself as a periodic waveform in the rest of the function, with the period in $\tau$ equal to the period of the oscillation in $t$. In other words, at particular values of the delay ($\tau$) the signal, being repetitive, is like earlier parts of itself. Autocorrelation accentuates rhythmicity by squaring these periodic aspects of the signal because of the multiplication process, whereas non-periodic aspects tend to cancel.

3. *Power spectrum.* One can go a stage further with the autocorrelation and obtain a spectral plot. This is done by subjecting the autocorrelation to a simplified Fourier transform. The spectrum obtained, however, is a

power spectrum rather than an amplitude spectrum, because the auto-correlation is fundamentally a squaring process. The autocorrelation is simply transformed to the power spectrum by

$$P(\omega) = \int_0^\infty r(\tau) \cos \omega\tau \, d\tau \qquad (4)$$

a formula equivalent to (1). One does not have to worry about the phase problem with this method, however, since the autocorrelation, with its maximum at $\tau = 0$, has effectively made all frequencies into cosine waves that add together at this value. Phase information is lost thereby, but it makes the operation of extracting the frequency components simpler, since one only has to probe with cosines. Examples of power spectra are shown in Figs 5 and 6.

As in all these signal analysis and statistical techniques one has to use a certain amount of discretion about the length of signal that one needs. The temptation is to use rather short lengths since computing time is valuable and each of these methods requires $n^2$ multiplications where $n$ is the number of samples in the signal being analysed. However, too short a length of record gives very large variance in the resulting functions, and this can easily obscure any patterns that one is looking for. One normal practice is to smooth the resulting functions by performing a type of running average of the data points. It is important not to use the simple running average, in which each point in the series being averaged contributes equally, since this imposes its own effect on the spectrum of the signal. Much better is to weight the signal by a method which ensures that points nearer to the point whose average is being plotted contribute more heavily than those further away.

An important advantage of being able to extract the periodicity of a signal by some such means as Fourier analysis, autocorrelation or power spectrum analysis, is that one is then in a position to start applying standard statistical tests to find out whether putative oscillations at the frequency of interest are significant. In Fig. 6, for instance, Sutton and Sykes (1967) were able to assess the significance of the 9 Hz tremor peak by calculating and plotting the standard error of the mean.

This has necessarily been a very brief introduction to methods for recognizing periodicities in a signal. Further information of a more detailed kind on these and other related signal analysis techniques can be found in Blackman and Tukey (1958), and Dern and Walsh (1963). The application of these methods is possible given access to computers, and techniques of this kind provide at least a basis for asserting confidently that a rhythm exists. At best they enable one to extract periodic signals which are not obvious in data viewed by the naked eye.

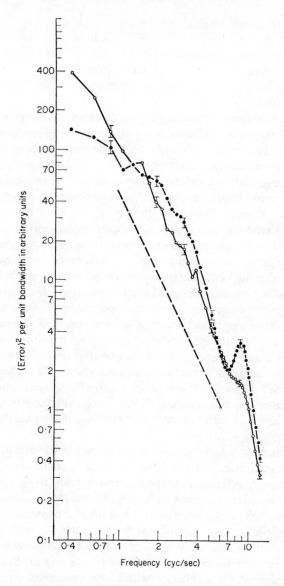

FIG. 6. Power spectrum of human muscular tremor in a tracking task with eyes open (●) showing the presence of a 9 Hz frequency in the tremor, and with eyes shut (○) showing the abolition of this particular oscillation. Vertical bars represent twice the standard error of the mean. The dashed line has a slope of 6 dB octave. Reproduced with permission from Sutton and Sykes (1967).

B. *Linear frequency analysis*

Frequency analysis is a method for deriving an input–output relationship and usually an analytical expression for a system. It does not depend on any inherent rhythmicity of the system being studied, but if there are oscillatory tendencies of a linear kind in the system then these methods will discover them.

The method consists of putting input signals consisting of sinusoidal waves into the system, and measuring the response. For linear systems, and for many that are not linear if the signal used is small and suitable approximations are used, the output signal is at the same frequency as the input test signal, but at a different amplitude and phase. For a series of inputs at different frequencies but at the same amplitude, graphs of the output amplitudes and phases are known as frequency response curves. The frequency response of a system is the exact equivalent of the spectrum of a signal. Whereas the spectrum says what frequencies a signal contains, the frequency response says what signals a system or network will allow to pass. A filter of the kind used in a speech spectrograph might for instance pass frequencies between 500 and 1000 Hz and when the speech contains components in that frequency band the output from that filter will be high.

Measuring the frequency response of systems has recently become popular because the mathematical simplicity of systems that can be regarded as approximately linear often allows one to find mathematical expressions (transfer functions) which represent the system. These methods have been used widely in tracking tasks to define models of the human operator (see for instance Licklider, 1960). They have also become a necessary part of the analysis of many other biological control systems (see Milhorn, 1966).

Temporal and spatial frequency responses have been used instead of the more usual measurement of responses to sudden, transient stimuli in the analysis of sensory systems. Perhaps the best known, most successful and the earliest of these experiments were those of Bekesy on the ear (collected in von Bekesy, 1960). He plotted the amplitude and phase of the sinusoidal oscillations of the basilar membrane response, when the oval window was driven sinusoidally with a mechanical vibrator. He found that the frequency response of the membrane was that of a filter. For any given point on the membrane there was a frequency which was preferentially passed. For frequencies lower than this there was a gentle attentuation as the frequency was diminished. On the high frequency side of the preferred point the attenuation slope was much steeper. He further found a whole family of tuning curves showing how a single frequency excited different parts of the membrane, indicating that while high frequencies are confined to the basal cochlea, low frequencies come to a maximum near the far (apical) end. It is this arrangement which on the "place theory" gives rise to our ability to discriminate different frequencies of auditory vibration, and which presumably is central to that important problem of human performance, the understanding of speech.

In some other biological systems where oscillations are important, linear frequency response methods are of less immediate value in analysing the system. Resonance phenomena, as found in the basilar membrane, tend not to occur in systems where the mode of oscillation is sharply non-linear. Particularly, as pointed out by Van der Pol (1940), they do not occur in relaxation oscillators. In these systems, rather than passing frequencies through them and encouraging them to oscillate with greater amplitude at one frequency than another, one resorts instead to attempting to determine the system characteristics by triggering the oscillations at certain phases of the cycle, or trying to entrain the rhythm with frequencies different from its normal one. It can easily be seen from this difference in approach that the response of an approximately linear oscillator is to oscillate at all frequencies with a greater or lesser amplitude and phase shift. A non-linear oscillator has a much more standardized mode of vibration, however; a periodic driving signal may or may not be capable of synchronizing it, and will not drastically increase its amplitude.

## C. *Entraining non-linear oscillations*

Many experiments on the analysis of rhythms in which some attempt has been made to analyse a non-linear oscillation have concentrated upon understanding that most obtrusive of rhythms both in human and animal performance, the diurnal cycle. Experiments on the effects of circadian variation on performance will be discussed in Chapters 2, 3 and 4, but it will be useful here to look in a slightly more abstract way at some of the experimental manipulations that one might wish to apply. These might be thought of as elements of a kind of non-linear frequency analysis.

1. *Phase shifting.* Most diurnal or circadian rhythms are synchronized by the daily pattern of light and dark, although in nature there are other diurnal patterns, e.g. environmental temperature, or rate of arrival of cosmic rays, which could conceivably be used to synchronize internal oscillators. There have been several well known experiments on men (see e.g. Kleitman, 1963), conducted in artificial lighting and constant temperature conditions, in which various diurnally fluctuating parameters, including body temperature, were measured as indices of adaptation to various light–dark régimes. Changes in phase of the lighting cycle are usually sufficient to change gradually the phase of endogenous rhythms; experiments in which this has not been achieved seem to have

been done in poorly controlled conditions. In animals it is nearly always possibly to reverse a diurnal rhythm, i.e. turn it 180° out of phase, by reversing the light–dark cycle. In general a complete reversal can take up to a week or 10 days though individual differences are wide. The maximum rate of phase change that seems to occur is about 60° (4 h) in a cycle when the maximum phase-shifting stimulus is applied.

Phase shifting has recently become particularly important when considering human performance because of the widespread growth of travel by jet aeroplane between time zones separated by several hours. The problem has recently been reviewed by Siegel, Gerathwohl and Mohler (1969). With phase shifts of 4 or more hours many travellers experience adverse effects, particularly when travel or arrival takes place during normal sleeping time. Diplomats and company executives should feel it necessary to take these disturbances of circadian rhythms into account when important decisions have to be made after travel across several time zones. Even more important perhaps is that airline companies take the effects into account, since their pilots may suffer the same deleterious consequences of phase-shifting. Siegel *et al.* (1969) suggest that aeroplane accidents are more likely to occur when pilots have suffered disruption of circadian rhythms.

2. *Frequency change.* An experiment less often performed consists of attempting to entrain rhythms on periods other than 24 h. Kleitman (1963) reports that it is possible to entrain the diurnal cycle in some subjects to a period of 18 h or to a period of 32 h. The same thing is achieved more easily with animals, no doubt because of the greater ease of standardizing conditions.

3. *Removal of synchronizing stimuli.* This is rather difficult to do in man, as in general even if the light pattern is removed he uses cues of other kinds, i.e. his alarm clock, conversation with others, etc. Exogenous rhythmic phenomena will be weakly coupled with endogenous oscillators, and as discussed above the salient feature of non-linear oscillators is to entrain even to very weak stimuli. Thus if, on an Arctic expedition, with constant light, a member of the team always goes to bed after making a routine daily radio communication, then this may be enough to keep the sleep rhythm in synchrony. Since this will be coupled to other rhythms by obvious links, e.g. one neither eats nor drinks when asleep, it is possible to imagine how the whole population of rhythms is kept synchronized. Complete removal of synchronizing stimuli has not always been totally successful in human subjects (see Kleitman, 1963), but in animals when all synchronizing stimuli are removed 2 things may happen.

A slight drift of phase from the solar cycle may occur. This is well seen in the running activity rhythm of the flying squirrel (De Coursey, 1960). The free running period in this animal was about 23·7 h. Alternatively in some conditions a rhythm will be reduced in amplitude or even abolished by removal of the light pattern.

4. *Dissociation of rhythms.* Some of the manoeuvres described above (1–3) are accompanied by different rates of phase shift in different circadian oscillators within the same individual. During a light reversal, for instance, some rhythmic parameter A may change its phase to the reversed pattern more quickly than another parameter B. This provides evidence that A and B, though normally running in synchrony, perhaps because of mutual coupling, or because of coupling to some third oscillator, do have their own private oscillation modes. Experiments of this kind make it seem as if many of the rhythmic parameters, e.g. body temperature, sleep, urine flow, etc. are each controlled by their own oscillator, which normally behaves synchronously with all the others. A further way of demonstrating independence is to level out one rhythm, and look to see what effect this has on another parameter to which it was closely coupled (e.g. Oatley, 1971). One might for instance suppose that a high correlation of body temperature with performance indicated a relationship of dependence of performance on temperature. If body temperature could be held constant experimentally, perhaps by supplying or removing heat from the surface of the body in such a way as to keep some central temperature measurement constant, then if performance were dependent on temperature it too should immediately become constant. If, as seems more likely, the performance rhythm continues, one could say that it must be driven by some other oscillator, perhaps arousal. Perhaps with a judicious use of drugs such as amphetamine, one could maintain arousal constant, and see whether the performance rhythm is independent of this. (For further discussion of these issues see Chapter 2.)

An important point is that whereas in many experiments one looks for strong (causal) connections between parameters, in the study of rhythms it is a mistake to suppose that this is the only kind of effective link between events. The repetition of events with only slight immediate effects but occurring in the right frequency range may have a powerful influence. This goes both for endogenous and exogenous influences.

5. *The optimal phase for stimulation.* With the kind of non-linear oscillation that is thought to underlie many biological periodicities it is entirely possible that if one wishes to make some drastic changes in a variable being influenced by a rhythm, one has to impose the right stimulus at the right

phase of the oscillation. Efficacy of drugs for instance varies according to the time of day at which they are given (see, e.g. Mills, 1966). Here again is a matter of principle in which practice differs fundamentally from work with steady state systems.

Following a theoretical analysis of diurnal rhythms of eclosion (emergence from pupae) in fruit flies, Winfree (1970) has found a phase of the cycle at which a pulse of light will completely abolish the rhythm. This sort of thing seems not to have been attempted on diurnal osciallations in man.

6. *The master clock?* Subjects of considerable theoretical interest are (a) what constitutes a physiological clock, (b) is there a master clock and (c) how are clocks coupled to the light–darkness cycle? Very little is known about (a). In subcellular timing systems it seems that metabolic control processes produce periodic changes of concentration of metabolic substances, and that these chemical oscillations within cells constitute one type of clock. Very little is known about the physiological processes underlying diurnal rhythms, particularly in man, although occasionally important components have been localized in animals.

It has been supposed that there might be some master clock and that in understanding diurnal rhythms in any animal we must locate it and then experiment upon its physiological properties. This, however, is probably a misunderstanding of the situation. Pittendrigh (1960) has proposed that we should "abandon the common current view that our problem is to isolate and analyse the endogenous rhythm or the internal clock". Instead of a single clock controlling or modulating the activities of what would otherwise be steadily maintained processes, we should, rather, expect that "the organism comprises a population of quasi-autonomous oscillatory systems" (p. 165). As was discussed above, the properties of nonlinear oscillators of the kind of which an organism may be composed is that they can entrain on one another to give the impression of being driven from a "master clock", and they can also become time locked with events going at higher or lower harmonics.

The way in which synchronization with the environment occurs is not much better understood than the physiological basis of the oscillations themselves. It is clear that some input from visual receptors is necessary, and it is plausible that a synchronizing signal might best be transmitted hormonally to all parts of the body. The possibility that the pineal gland, and its light-induced secretion of melatonin, may be the route for exogenous synchronization is currently attracting attention (Wurtman, Axelrod and Kelly, 1968), though this mechanism is by no means firmly established.

## VI. Conclusions

Analysis of many rhythms in man is of course more difficult than in animals. Experimental manipulations such as phase-shifting and synchronization to different day-lengths are very expensive if done properly on man, so that often rather incomplete and unsatisfactory data have to be accepted as all that can conceivably be acquired in many situations. It is also difficult in most circumstances to collect data in man for more than a very few cycles when rhythms have a period of a day or more. This too militates against the use of some of the mathematical tools that are mentioned here. Many of them, e.g. spectral analysis, require at least 10, and preferably many more cycles before they can sensibly be employed. Nevertheless, armed with some of these mathematical ideas and some of the more provocative results of animal experimentation, we can begin to make sense of human performance rhythms.

Rhythms may occur as an essential manifestation of living processes. If the view put forward here is correct then we would expect many processes of interest in the study of human performance to be based on oscillation, and the oscillation may be fundamental to understanding, rather than a complicating side issue.

In applied problems of human performance one needs to be aware of rhythms and their properties in order to take advantage of them, or at least not be put at a disadvantage because of them. If one wishes to manipulate a rhythmic process, there is probably an optimal phase at which to do so. Rhythmicities within the organism are probably communicated by weak coupling between more or less autonomous oscillators, and, in the same way, in order to affect a rhythm from outside, even weak coupling to a source of periodic stimulation may be enough.

We are so familiar with some of the main rhythms of life, sleeping and waking, menstruation etc. that we organize our personal activities to fit in with them. It is probably not the case however that social organizations take these things into account to any greater extent than having employment organized round an 8-h working day. If we understood more about the effects of rhythms and how to manipulate them, then there might well be opportunity for increases both in efficiency of work and in enjoyment of leisure.

Cycles of both longer and shorter periods than 24 h, can easily go unnoticed. Even when the behavioural effects of such cycles are gross and bizarre, as in some of the major types of psychotic illness which recur within a period of weeks or months, few have seen the rhythmicity of the illness as important, except for the purposes of diagnosis. Richter (1965), however, has pointed out the wide incidence and relevance of rhythmic phenomena in medicine. No doubt several of these rhythmic disturbances have a basis in normal physiological rhythms which affect performance in more subtle ways. The curious interaction of endogenous factors and external events in precipitating depression, for instance, is comprehensible in terms of an oscillator-structure of the organism which we have discussed, but possibly not easily otherwise. If we understood more about the rhythms of the body and brain we might be able to decide more sensibly when to work, when to go on holiday, when to do the things that require concentration, and even how to raise ourselves out of sloth and inactivity.

## References

Akert, K., Koella, W. P. and Hess, R. (1952). *Amer. J. Physiol.* **163**, 260–267.

Alpern, M. (1958). *J. Opt. Soc. Amer.* **48**, 193–197.

Aschoff, J. (ed.) (1965). "Circadian Clocks". North-Holland Publ. Co., Amsterdam.

Bekesy, G. von (1960). "Experiments in Hearing". McGraw-Hill, New York.

Betz, A. and Chance, B. (1965). *Arch. Biochem. Biophys.* **109**, 585–594.

Blackman, R. B. and Tukey, J. W. (1958). "The Measurement of Power Spectra". Wiley, New York.

Broadbent, D. E. (1958). "Perception and Communication". Pergamon, London.

Broadbent, D. E. and Gregory, M. (1967). *Proc. Roy. Soc. Ser. B.* **168**, 181–193.

Bruce, V. G. (1965). *In* "Circadian Clocks" (J. Aschoff, ed.), pp. 125–138. North-Holland Publ. Co., Amsterdam.

Bunning, E. (1964). "The Physiological Clock". Springer-Verlag, Berlin.

Chance, B., Hess, B., and Betz, A. (1964). *Biochem. Biophys. Res. Commun.* **16**, 182–189.

*Cold Spring Harbor Symp. Quant. Biol.* **25**, 1960.

Cowan, J. D. (1969). *In* "Mathematical Questions in Biology" (M. Gerstenhaber, ed.), American Mathematical Society, Providence.

Craik, K. J. W. (1943). "The Nature of Explanation". Cambridge University Press, London and Cambridge.

Craik, K. J. W. (1947). *Brit. J. Psychol.* **38**, 56–61.

De Coursey, P. (1960). *Cold Spring Harbor Symp. Quant. Biol.* **25**, 49–55.

Dern, H. and Walsh, J. B. (1963). *In* "Physical Techniques in Biological Research" (W. L. Nastuk, ed.), Vol. VIB, pp. 99–218. Academic Press, London and New York.

Eccles, J. C. (1969). *In* "Information Processing in the Nervous System" (K. C. Liebovic, ed.), pp. 245–269. Springer-Verlag, Berlin.

Efron, R. (1967). *Ann. N.Y. Acad. Sci.* **138**, 713–729.

Fogel, L. J. (1963). "Bio-technology: Concepts and Application". Prentice-Hall, Englewood Cliffs, New Jersey.

Goodwin, B. C. (1963). "Temporal Organization in Cells". Academic Press, New York and London.

Goodwin, B. C. (1969). *Symp. Soc. Gen. Microbiol.* **19**, 223–236.

Hamburger, V. and Balaban, M. (1963). *Develop. Biol.* **7**, 533–545.

Hammond, P. H., Merton, P. A. and Sutton, G. G. (1956). *Brit. Med. Bull.* **12**, 214–218.

Hess, R. Jr., Koella, W. P. and Akert, K. (1953). *Electroencephalogr. Clin. Neurophysiol.* **5,** 75–90.
Hoagland, H. (1963). *Cold Spring Harbor Symp. Quant. Biol.* **4,** 267.
Horn, G. (1967). *Nature (London)* **215,** 707–711.
Iberall, A. S. (1969). *In* "Towards a Theoretical Biology" (C. H. Waddington, ed.), Vol. II, pp. 166–177. Edinburgh University Press, Edinburgh.
Kauffman, S. A. (1969). *J. Theor. Biol.,* **22,** 437–467.
Kleitman, N. (1963). "Sleep and Wakefulness" (Rev. Ed.). University of Chicago Press.
Lashley, K. S. (1951). *In* "Cerebral Mechanisms in Behaviour" (L. A. Jeffress, ed.), pp. 112–136. Wiley, New York.
Latour, P. L. (1967). *Acta Psychol.* **27,** 341–348.
Licklider, J. C. R. (1960). *In* "Developments in Mathematical Psychology" (D. Luce, ed.), pp. 168–279. Free Press of Glencoe, Illinois, U.S.A.
Lobban, M. C. (1965). *In* "Circadian Clocks" (J. Aschoff, ed.), pp. 219–227. North-Holland Publ. Co., Amsterdam.
Merton, P. A., Morton, H. B. and Rashbass, C. (1967). *Nature (London)* **216,** 583–584.
Milhorn, T. (1966). "The Application of Control Theory to Physiological Systems". W. B. Saunders, Philadelphia.
Mills, J. N. (1966). *Physiol. Rev.* **46,** 128–171.
Minorsky, N. (1962). "Nonlinear Oscillations". van Nostrand, New York.
Neisser, U. (1967). "Cognitive Psychology". Appleton Century Crofts, New York.
Newman, E. A. and Evans, C. R. (1965). *Nature (London)* **206,** 534.
Oatley, K. (1967). *Med. Biol. Eng.* **5,** 225–237.
Oatley, K. (1971). *Nature (London)* **229,** 494–496.
Othmer, E., Hayden, M. P. and Segelbaum, R. (1969). *Science* **164,** 447–449.
Pattee, H. H. (1970). *In* "Towards a Theoretical Biology" (C. H. Waddington, ed.), Vol III, pp. 117–136. Edinburgh University Press, Edinburgh.
Pittendrigh, C. S. (1960). *Cold Spring Habor Symp. Quant. Biol.* **25,** 159–184.
Priban, I. P. and Fincham, W. F. (1965). *Nature (London)* **208,** 339–343.
Prigogine, I. (1967). "Introduction to the Thermodynamics of Irreversible Processes" (3rd Ed.). Interscience Publishers, J. Wiley and Sons, New York.
Pringle, J. W. S. (1951). *B. .our* **3,** 174–215.
Richter, C. P. (1965). "Biological Clocks in Medicine and Psychiatry". C. C. Thomas, Springfield, Illinois.
Ritow, I. (1963). "Capsule Calculus". Macmillan, London.
Shallice, T. (1964). *Brit. J. Stat. Psychol.* **17,** 113–135.
Siegel, P. V., Gerathwohl, S. J. and Mohler, S. R. (1969). *Science* **164,** 1246–1255.
Sokolov, E. N. (1960). *In* "Central Nervous System and Behaviour" (M. A. B. Brazier, ed.), pp. 187–276. Josiah Macy Jrn. Foundation, New York.
Stanten, S. F. and Stark, L. A. (1966). *IEEE Trans. Bio-Med. Eng. B.M.E.* **13,** 140–152.
Stark, L. (1959). *Proc. I.R.E.* **47,** 1925–1939.
Stroud, J. M. (1955). *In* "Information Theory in Psychology" (H. Quastler, ed.), pp. 174–207. Free Press, Glencoe, Illinois.
Sutton, G. G. and Sykes, K. (1967). *J. Physiol (London)* **190,** 281–293.
Van der Pol, B. (1940). *Acta Med. Scand. Suppl.* **108,** 76–87.

Venables, P. H. (1960). *Brit. J. Psychol.* **51,** 37–43.
Webb, W. B. (1968). "Sleep: an Experimental Approach". Macmillan, New York.
Welford, A. T. (1952). *Brit. J. Psychol.* **43,** 2–19.
White, C. (1963). *Psychol. Monog.* **77,** No. 12, Whole No. 575.
Wilde, R. W. and Westcott, J. H. (1962). *Automatica* **1,** 5–19.
Winfree, A. T. (1970). *In* "Lectures on Mathematics in the Life Sciences" (M. Gerstenhaber, ed.), Vol II, pp. 111–150. American Mathematical Society, Providence.
Wurtman, R. J., Axelrod, J. and Kelly, E. D. (1968). "The Pineal". Academic Press, London and New York.
Young, L. R. and Stark, L. (1965). "Biological control systems—a critical review and evaluation: Developments in manual control". *N.A.S.A. Report CR-190.*

CHAPTER 2

# Circadian Variations in Mental Efficiency

W. P. COLQUHOUN

*MRC Unit of Applied Psychology, Laboratory of Experimental Psychology, University of Sussex, Brighton, Sussex, England*

## I. Introduction

The average man sleeps for a single stretch of some $7\frac{1}{2}$ h out of each 24-h day. This leaves over 16 consecutive hours in which, apart from the time devoted to activities necessary for survival and for attention to personal hygiene, he is available to perform such work as is demanded of him by the society in which he lives. In industrialized communities this work takes up some 8 of these 16 h, and usually commences within 1 or 2 h of awakening from sleep, the exact interval often depending largely on the distance between home and workplace. Since the preferred hours of sleeping are, for reasons probably inherited from our remotest ancestors, approximately coincident with the hours of darkness, the average man's working day therefore starts between 07.00 and 09.00 in the morning. For administrative convenience the work is usually performed in a single spell, with a suitable break for eating about half way. Thus the working day ends somewhere between 16.00 and 18.00 in the evening. Between the end of work and the start of sleep, "leisure activities" complete a cycle which repeats itself with little variation on each of at least 5 days of the week.

The most obvious division of this 24-h cycle is into 2 phases—the sleep phase, and the wakeful or "active" phase. Physiologists have known for some time, however, that this 2-state description of the cycle is, biologically speaking, considerably over-simplified. Although it is true that daily variations in heart rate, blood pressure, respiration rate, body temperature and urinary excretion are approximately synchronized so that they all show high levels during the daylight hours and are at their lowest in the hours of darkness, none of these processes exhibits that simple oscillation between 2 levels which would be consonant with a 2-state hypothesis. On the contrary, it can readily be shown that variation is **continuous** throughout the 24-h period; it is only the "peaks" and the "troughs" of the circadian rhythms which actually tend to coincide with light and darkness respectively. Thus for the greater part of the 24-h period many, if not all of the physiological processes of the organism are in a state of continuous change (cf. Mills, 1966).

One of the most readily observable of these physiological processes is the internal temperature of the body. Homeostatic mechanisms ensure that variation in this temperature in the healthy individual does not exceed more than 1 or 2 degrees in all but the most extreme environmental conditions. Nevertheless, within this "permitted" range systematic daily changes do occur. Throughout each 24-h period the oral temperature, for example, fluctuates about a mean that is usually well below the level of 98.4° F that is often regarded as "normal".

This circadian variation in body temperature is illustrated by readings taken from a group of 70 young men (Naval ratings) during the course of a 24-h period in which they followed a sequence of activities similar to our "average industrial man". The temperatures were recorded by clinical thermometers inserted sub-lingually for 3 min at hourly intervals during the waking period, and at 2-h intervals during sleep (the subjects were roused gently on these occasions). Thus 20 readings were taken from each man. This procedure was repeated for a second 24-h period, after which both sets of readings were summed over all 70 men, and 3-point weighted moving averages computed for each time of day or night.

The range of variation in the smoothed curve thus derived was 1.19° F, with a peak reading of 98.35° at 20.00 h, and a trough of 97.16° at 05.00 (see Fig. 1). The fall from maximum to minimum values therefore took place over a period of some 9 h, but the rise extended over a period almost twice as long as this. Although the fall was approximately linear

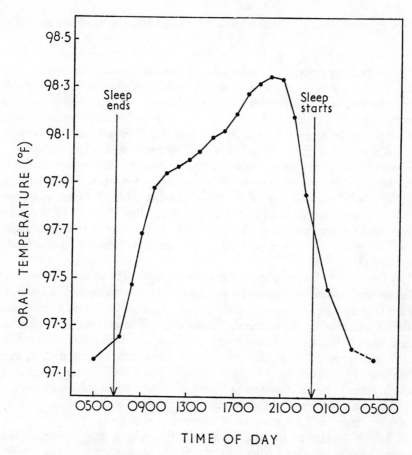

FIG. 1. Circadian rhythm of oral temperature in a group of 70 young men.

over its major part, the rise was clearly not, since 60% of it had occurred by 10.00, i.e., in only one-third of the total time. Thereafter, for a period of some 10 h between 10.00 and 20.00, the temperature rise was relatively slow.

Although this later period of relative stability embraces some 60% of our average man's "waking day", it does not include the first 1–3 h of his normal **working** day. During the period 07.00–10.00 the temperature in the group of subjects studied increased at an average rate of 0·24° per h, which is roughly 5 times the rate recorded over the succeeding hours. In so far as the routine of these young man was representative of industrial workers in general (they got up at 06.30, and worked from 08.00 to 16.30), this means that for the first 2 h of the working day temperature is rising very rapidly, and that even by the end of the working period it has still not reached an allegedly "normal" level (though it has in fact exceeded the mid-point of the total daily variation at this time).

Since body temperature varies in such a striking way throughout the day, and since it is known that many other physiological processes exhibit similarly marked circadian changes, it appears more than likely that the functioning of the brain will also vary, in that its operation must clearly be supported by some at least of these processes. There are therefore strong grounds for expecting that the speed and accuracy with which all manner of activities involving nervous function are executed will fluctuate throughout the day. The extent to which this variation affects actual performance "on the job" should theoretically be determinable by observing the efficiency with which people carry out their allotted tasks at different times during the working day. However, the question immediately arises at this point—how is "efficiency" at these tasks to be actually measured?

At first sight, the answer to this question appears obvious. Assuming that the task is one in which the production can be measured in units, it is a relatively simple matter to arrange for, say, hourly counts to be made of the number of units produced. "Work-curves" based on counts such as these have been obtained in a number of studies on industrial fatigue in relation to type of work, length of shift, frequency of rest pauses, etc. (Viteles, 1932). Such curves have recently been used by organizations which provide piped music to factories, as a somewhat dubious basis for varying the content of the programme at different points in the shift in order to "iron out" the observed fluctuations in output.

While work-curves of this kind tell us much of interest about the **overall way** in which a particular operation is being carried out, it is most dangerous to assume that because, for example, more units are produced between 10.00 and 11.00 than between 07.00 and 08.00, the worker has necessarily become genuinely more **efficient** at performing his particular task during the morning. The factors in a typical manufacturing plant which could affect an hourly output count are almost too numerous to list. Provided they could be recorded, automatic compensation could be made for interruptions in the flow of raw material to the operator, machine stoppages, or rest-breaks (official or unofficial). However, the **psychological** effects of these pauses on output would, of course, be unknown, as would the influence of any alteration in ambient conditions such as heat, noise and illumination-level. When it comes to the effect of what might in general be termed "social" factors such as the individual's interactions with his supervisor and his workmates, whether he is being paid on a time- or piece-rate basis, and the extent to which he is attempting to conform to acceptable "norms" of production, it will be seen that the measurement of "true" efficiency in the real-life situation presents some almost insuperable obstacles.

For these reasons the emphasis in the present discussion of circadian rhythms in performance will be mainly on results from **experimental** investigations with carefully controlled laboratory tasks. Here nearly all of the unwanted variables listed above are eliminated, and we can be considerably more certain that what has been measured is the genuine efficiency-level of the subject at a particular time of day. Since, as is usually the case, the laboratory task is novel to the subject, the very real difficulty of replacing the socio-economic incentives of the factory with adequate motivation to perform well in the experimental situation is at least partly solved by the automatic arousal of interest in what has to be done. Task novelty provides the additional advantage that, by definition, there are no "norms" of performance with which the subject can compare his own efficiency. Against these advantages must be set the cost of a substantial loss of realism, but this would appear to be inevitable.

Most of the tasks which are used in the study of circadian effects are what is commonly described as "mental". This choice is sometimes made for reasons of practical expediency, but there is in truth some justification for concentrating on this kind of activity. Apart from the fact that it is often simpler to **construct, administer,** and **score** a task which involves relatively little motor activity, performance at mental tasks is probably less likely to be affected by **transfer effects** from specific skills acquired in the course of normal everyday life. On the other hand it is considerably easier to make a mental task appear more meaningful to the subject (e.g. by describing a signal detection task as being analogous to a product inspection operation) than it is a motor one, without going to relatively extreme lengths of simulation. It might also be argued that the greater the mental content of a task the greater will be its **sensitivity to variation in nervous function**, though the grounds for this reasoning are perhaps somewhat tenuous. What is not in doubt is that, in actual industrial operations, the advance of automation is resulting in a decline in the number of jobs which involve motor skills, and a corresponding increase in those requiring almost pure mental activity; thus in terms of the applicability of research findings to our average industrial worker, experiments with predominantly mental tasks are clearly more relevant.

The majority of these research findings fall into 2 groups

(a) those concerned with variations in performance throughout the **normal "waking day"**, i.e., experiments in which the subject had his usual hours of sleep at his customary time and

(b) those in which efficiency has been observed **"round the clock"**, i.e., experiments during which some at least of the measurements were taken in "night" hours.

The first group of findings should have application to the question of real-life efficiency not only during the "average" working day, but also during both "early" and "late" turns in organizations where a 2-turn day shift system is in operation. These early and late turns are typically held from 06.00 to 14.00, and 14.00 to 22.00 respectively; they thus involve very little, if any, disturbance of normal sleep hours. On the other hand, such disturbance is an automatic corollary of experiments in the second group, and this introduces complicating factors which affect the interpretation put upon the results obtained. However, these results are of particular relevance to work efficiency on "night" shifts, and since the problem of organizing these night shifts is rapidly increasing in urgency as 24-h operation of plant machinery becomes an economic necessity, we shall be devoting special attention to studies in this second group, and also to a unique class of investigations which specifically set out to simulate the kinds of work-rest schedule that shift-workers have to follow in different types of system, in order to determine the changes in performance that occur as a result of adjusting to abnormal routines.

## II. "Waking Day" Studies

"That the individual's capacity for doing mental or physical work is not the same throughout the waking period has been known for a long time." Thus Kleitman (1963) sums up what is probably a widely held opinion. The trouble is, although most people would accept that their efficiency varies somewhat during the day, few would agree entirely on which times were most propitious, and which were worst, for carrying out their assigned tasks. This is no doubt due partly to differences in the kind of work that particular individuals have to perform, but it may also reflect intrinsic differences in the phasing of the circadian fluctuation in those physiological processes which influence nervous activity. There is good evidence that individual differences in the phasing of at least the body-temperature rhythm do in fact exist, and experiments have shown that associated differences in the phasing of performance curves can in some cases be demonstrated by laboratory tests. However, here we are concerned with the question whether there is any **general** pattern of variation in performance functions during the waking day, on to which such individual differences are superimposed.

Early research findings in this area are difficult to interpret because of the small numbers of subjects usually employed, and sometimes, because of failure to control for possible confounding effects from practice or "learning" on successive trials. In fact Kleitman (1963), who must be regarded as the leading authority on the present subject, does not appear to give very much weight to most of this early work, preferring to present, in detail, the results from his own carefully controlled studies. Kleitman appears to have had good justification for doing this. In a review published in 1934, Freeman and Hovland prepared a table of the experimental evidence available at that date on various aspects of daily variation in "sensory", "motor", and "mental" performance. They divided the performance curves into 4 classes: a continuous rise during the whole period of observation; a continuous fall; a morning rise and an afternoon fall; and a morning fall and an afternoon rise. This is tantamount to saying that there is no consistent pattern whatsoever, but Kleitman clearly does not believe this to be the case. He found that all the curves he and his co-workers obtained fitted the third class distinguished by Freeman and Hovland, with a peak of performance in the middle of the waking period, i.e., somewhere during the afternoon. As will be shown later, results from our own laboratory indicate that peak performance at most of the tasks we have studied tends to occur in the evening rather than the afternoon, if and when a daily trend is observable (the "afternoon" period in our studies was commonly associated with a relatively **low** level of performance; possible reasons for this will be discussed).

Although our results may differ from Kleitman's in respect of the time of maximum efficiency, we, like him, also have little doubt that there is a recognizable general pattern in the fluctuations of performance levels during the waking day. Kleitman relates this pattern to the 24-h sleep-wakefulness rhythm, and, by taking body temperature as an index of this rhythm, seeks to demonstrate that temperature and performance variations are in fact coincident during the entire course of the day. In the case of tasks involving motor activity he implies that this parallelism is mediated by the underlying relationship between body temperature and muscular tonus. Where the task is predominantly mental in character he suggests two possible interpretations of the co-variation of efficiency and temperatures: "Assuming that the effect of body temperature indicates that one is dealing with a chemical phenomenon (then) either (a) mental processes represent chemical reactions in themselves or (b) the speed of thinking depends upon the level of metabolic activity of the cells of the cerebral cortex, and by raising the latter through an increase in body temperature, one indirectly speeds up the thought process" (Kleitman, 1963, p. 160). He concludes that "most of the curves of performance can be brought into line with the known 24-h body-temperature curves, allowing for individual skewing of the curves towards an earlier or later, rather than a mid-afternoon, peak" (p. 161).

Although our own researches suggest that this last statement requires some slight qualification, we would not dispute the fact that diurnal fluctuations in performance (where observed) are **in general** associated with concomitant variations in body temperature. However, certain specific discrepancies observed in our own experiments seem to us to be of sufficient importance to question the value, at the present time, of putting forward hypotheses concerning possible linkages, chemical or otherwise, between the two. We will therefore confine ourselves, for the moment, to a discussion of the observed facts, and Kleitman's own studies provided a convenient starting point for this.

## A. *Temperature and efficiency*

In an early series of trials with 6 subjects, Kleitman used a number of tests, some of which, as he says, required "more cerebral activity" than others. These tests were (a) card dealing (b) card sorting (c) "mirror drawing" (d) nonsense-syllable copying (e) simple code transcription (f) multiplication (g) hand-steadiness and (h) body-sway. They were performed 5 times daily: immediately upon getting up, 1 h later, just before the noon meal, just before the evening meal, and just before going to bed. In the "mental" tests (a–f) the recorded scores were the time taken to complete the task, and, also (in tests c–f) the number of errors made. In summarizing the findings from these experiments, Kleitman states that "There was an agreement in the results obtained on the same test in different subjects. The curves of both speed and accuracy of performance showed a well-marked rhythm, with minima in the morning and late at night and a maximum in the middle of the day" (Kleitman, 1963, p. 151). He found that the temperature curve was parallel to that of performance except during the period between 12.00 and 18.00; between these 2 times performance declined somewhat, whereas the composite body temperature was still rising slightly. However, in the case of 1 subject who did 10 tests daily instead of 5, a graph (p. 152) shows that there was a very close correspondence between performance and temperature curves throughout the day, and Kleitman adds that "The swaying (ataxia) tests, as well as hand steadiness, showed a similar parallelism between the body temperature curves and those of performance" (p. 153).

Convinced, by the results of these experiments, of the existence of a definite relationship between body temperature and task efficiency, Kleitman and his colleagues went on to study performances which were, as he put it "purely sensorimotor and sensory-mental-motor". These two categories were represented by simple reaction time ($RT$) and choice $RT$ respectively. Tests were carried out, again on 6 subjects, at various times between 07.00 and 23.00, using both visual and auditory stimuli (Kleitman,

Titelbaum and Feiveson, 1938). It was concluded from this investigation that performance exhibited a diurnal curve, consisting of a progressive decrease in $RT$ during the morning and early afternoon and a rise in the late afternoon and evening, and that this variation was "frequently related to the diurnal temperature curve". A sample curve of the visual choice results for 1 subject is given in support of this contention, together with a table of the full results from this subject and from 3 others used in the study. In these other subjects a relationship is also apparent; no mention is made of the results obtained from the remaining 2 subjects.

On the evidence obtained in these trials, Kleitman concludes that "there is probably no $RT$ curve independent of the temperature. On the contrary, it would appear that $RT$ was always connected with the body temperature, which, whenever it changed, was accompanied by a change in the opposite direction in $RT$" (Kleitman, 1963, p. 154). Bearing in mind that this conclusion is based on a study which involved only 6 subjects, we ourselves feel that further research is needed before it can be accepted without reservation that changes in reaction time during the waking period are **always** associated with variations in temperature.

One aspect of Kleitman's investigations which should be emphasized is the fact that observations were continued over 20 days or more. Because of this, it was possible to examine the diurnal performance fluctuations at different stages of practice at any particular task. Apparently the daily curves were not affected by such practice. In the case of $RT$, for example, "There was for a time a gradual improvement in $RT$ scores, but this improvement was not related to the influence of body temperature, which manifested itself at different levels of achievement" (Kleitman, 1963, p. 154). This demonstration that the "time of day" effect is independent of the degree of familiarity with the task is most important, not least because it goes a long way towards refuting the argument that laboratory results cannot readily be applied to real-life situations on account of the extensive practice "on the job" that the factory or office worker has had.

Another interesting finding in Kleitman's reaction-time study was that the effect of temperature was more marked in the choice $RT$ situation than in the simple $RT$ case. Kleitman considers that the significance of this result lies in the fact that a higher degree of mental activity is required of the subject in the "choice" task. The implication here is that mental processes are more sensitive to the kind of variation under study. Thus the greater the mental content of a task, the more pronounced should be the fluctuation in efficiency at performing it through the waking day.

Although the mental content of a given task is not something which is readily quantifiable, it would seem reasonable to describe the activity

involved, for instance, in estimating the passage of time to be almost
entirely mental. We might therefore expect that such estimates would be
particularly sensitive to time of day effects, in so far as it is accepted that
these are mediated through temperature changes. In fact, it has been
frequently demonstrated that **induced** temperature changes do give rise
to marked alterations in the judgment of particular time intervals
(Hoagland, 1933; Kleber, Lhamon and Goldstone, 1963; Baddeley, 1966;
Fox *et al.*, 1967; Lockhart, 1967), and these results are generally interpreted as supporting the existence of a "chemical clock". However, the
evidence on the effects of **naturally occurring** changes in temperature on
the ability to estimate time is conflicting. Thus whereas Francois (1927)
and Pfaff (1968) both obtained results which implied that the "clock"
accelerated as temperature rose during the day, Thor's (1962a) findings
seem to point to the **opposite** conclusion; and Blake (1967) found no
consistent trend at all.

Perhaps we should not take "mental" to mean simply "relative absence
of motor activity", but rather, something more akin to the commonly
used phrase "task complexity". On this dimension (ill-defined though it is)
Kleitman claims to have shown, by application of the Van't Hoff-
Arrhenius equation (which yields a value "$\mu$" representing the critical
thermal increment in calories per gram mol of a particular reacting
system) to the fitted curves relating performance data to temperature in
his first study, that "in general, simpler tasks gave low $\mu$ values, and more
complicated acts, higher values" (Kleitman, 1963, p. 159). In the *RT* study,
simple reaction times were found to give smaller values of $\mu$ than choice
reaction times.

The Arrhenius equation has been used mainly to assess the relative
speeds of purely "biological" processes in relation to temperature (cf.
Crozier, 1926). Thus its applicability to performance data is open to question. Perhaps for this reason it does not seem to have been employed by
other investigators working with behavioural measures in the present
field. Unfortunately, in the absence of any agreed alternative technique for
equating the raw data obtained from different tasks, it is difficult to draw
conclusions about the relative sensitivity of various kinds of performances
to diurnal effects. Blake (1967), whose study included 8 different tests,
does not attempt to do so, perferring simply to give the level of statistical
significance associated with the overall differences in mean performance
levels at 5 times of day recorded from samples of 25 or 30 subjects. Comparison of these significance levels indicates that tests of card-sorting,
letter cancellation, and addition were more reliable indicants of time of day
effects than were tests of vigilance, serial-reaction, and digit span, despite
the fact that greater observed proportional variations in score tended to be

found in the latter group. Interpretation of Blake's results will not be attempted here, since they are discussed in detail in the following chapter; however, it is of interest to note that Blake found a pronounced "post-lunch" dip in the level of performance at **all** tests where efficiency otherwise rose significantly throughout the day.

### B. *The "post-lunch" phenomenon*

This "post-lunch" dip is of special interest, since as there is no corresponding drop in the temperature level after lunch (cf. Fig. 1) its occurrence would appear to refute Kleitman's thesis that temperature change is the **sole** determinant of diurnal fluctuation in performance. Kleitman himself does not mention having observed any such dip in his own results, but he does cite 3 earlier experiments in which this was found (Gates, 1916; Lehmann, 1953; Rutenfranz and Helbruegge, 1957). According to Kleitman the dip "may be related to a 'let-down' experienced after heavy noontime meals" (Kleitman, 1963, p. 158). Whether this "let-down" is due to digestion of the meal itself is by no means clear. M. J. F. Blake (unpublished data) has carried out 2 experiments in which the time of the "mid-day" meal was deliberately varied, in order to determine whether the actual presence of food in the stomach is a necessary condition for the occurrence of the effect. In both experiments the task was simple letter-cancellation, where a definite "dip" in speed of work is known to occur at 13.00 following a meal at 12.00 (Blake, 1967).

In the first study, subjects were tested at 13.00 on 3 separate occasions, having their lunch either at the usual time (12.00), or after the test (14.00), or an hour earlier than usual (11.00). Three groups of 12 subjects were employed, each group receiving the "treatments" in a different order. Mean output scores (number of letters checked in 30 min) over all 36 subjects for the 3 "time of lunch" conditions were: 12.00 lunch, 1311; 14.00 lunch, 1313; 11.00 lunch, 1268.

It is clear from the results for "12.00" lunch and "14.00" lunch that performance at 13.00 is essentially identical whether or not the subjects have had their usual meal beforehand. Thus on the face of it the actual presence of food in the stomach would appear to be ruled out as a factor contributing to the dip in the performance curve 1 h after lunch. However, it is of course possible that changes in physiological processes conditioned to a meal at 12.00 would still occur if that meal was withheld on only a single occasion, and that it is **these** changes that are affecting performance. A conclusive test of this could only be made by withholding lunch until after the test over a number of days or even weeks, to allow time for extinction of these conditioned responses.

An unexpected finding in this study was the lowered performance level in the "11.00" lunch group. This suggests the possibility that the greatest dip in the "whole-day" performance curve would appear, in "normal" circumstances (i.e., lunch at 12.00), at 14.00 rather than 13.00. Further evidence on this point was obtained in the second study by Blake, which, however, produced some equivocal findings.

In this study there were 4 conditions: (1) lunch 11.00, test 12.00; (2) lunch 11.00, test 13.00; (3) lunch 12.00, test 13.00; and (4) lunch 12.00, test 14.00. Note that conditions (2) and (3) corresponded to the 2 "lunch before test" conditions of the first study. Four groups of 5 subjects were tested, each receiving the "treatments" in a different order. Mean output scores for the 4 conditions were as follows: Condition 1, 1377; Condition 2, 1419; Condition 3, 1426; Condition 4, 1365.

Interpretation of these results is difficult, since the grouping of the conditions into 2 sub-classes evident in the scores does not correspond with any obvious classification, e.g. "2-h delay/1-h delay", or "early lunch/ late lunch". The 2 high output condition ((2) and (3)) have in common the actual test time (13.00), but the fact that the scores were almost identical in the 2 conditions despite the difference in the preceding post-meal interval throws doubt on the validity of the effect of this interval that appeared in the first experiment. On the other hand, the depression of performance in Condition (4) relative to that in Condition (3) is in line with the suggestion arising from the results of the first study that the dip in the "whole-day" curve should be greater at 14.00 than 13.00.

Clearly, much further research is required on the effect of meal **time** and post-meal **interval** on performance before any definite conclusions can be made about the relationship (if any) between these factors and the "post-lunch" dip in the efficiency curve. One thing appears quite definite, however: the dip is **not** accompanied by a measurable drop in temperature level.

## C. *Arousal and performance*

In the absence of any obvious explanation of the post-lunch phenomenon, we must seek for clues as to its cause in the subjective impressions of people who exhibit a drop in their performance at this time of day. When these impressions are analysed, it is clear that the one feeling that stands out above all others is that of **sleepiness**. It is usual on the part of respondents to ascribe this feeling to the digestion of the meal itself, but if digestive processes were in fact in some way responsible one would expect similar reports after breakfast, tea and dinner, and these do not normally occur (typically one is, if anything, **less** sleepy after these meals than before, except perhaps when dinner is taken very late).

If we now equate "sleepiness" with "arousal" (or rather the lack of it) as Corcoran (1962) has proposed, we can see how not only the post-lunch dip, but all other points on the waking-day curve of performance efficiency can be related to the state of arousal or "activation" at the time of testing (for a discussion of the concept of arousal, see Duffy, 1962).

It must be postulated that the **general** level of sleepiness falls (i.e., arousal rises) during the waking day, to reach a minimum somewhere in the evening. Following this the level rises again until a state of actual sleep is achieved. Additionally, we must suppose that a **temporary** increase in sleepiness-level (decrease in arousal) occurs at or about the time at which it is customary (perhaps not simply by coincidence) to take lunch. This temporary increase in sleepiness could in one sense be described as part of a short-lived secondary cycle superimposed on the overall 24-h sleep-wakefulness rhythm. This cycle may have originally served some biological need that no longer exists in adult industrial man; in fact Kleitman (1963) holds that the post-lunch "let-down" may reflect the "rest" phase of an underlying rest-activity periodicity of some 80–90 min duration which is **more** basic than the 24-h rhythm itself. A short-term periodicity of this order can be seen in the REM-EEG cycles associated with dreaming in actual sleep, and Kleitman suggests that it may persist during the waking period also, manifesting itself in recurrent fluctuations in alertness.

Whether the post-lunch dip in performance reflects a single episode of relatively reduced arousal occurring during a period in which the level is otherwise steadily increasing, or (as Kleitman suggests) one of several such phases occurring during the day, it is clear that because of its un-doubted existence it follows that we cannot take body temperature as an index of the level of arousal throughout the day except at times where changes in the former happen to coincide with fluctuations in the latter. Thus, assuming the theory that arousal determines efficiency to be correct, it is only at such times that we would expect performance and temperature variations to be correlated. However, since

(a) the body-temperature curve does in any case follow the **overall** "sleepiness" or "fatigue" rhythm relatively closely (Murray, Williams, and Lubin, 1958),

(b) the post-lunch period is frequently omitted in time of day studies, and

(c) measures of performance are not normally sensitive enough to reflect short-lived arousal fluctuations (supposing that these do in fact occur), it is not altogether surprising that such correlations are often found. Nevertheless, a great deal of caution should be exercised in inferring from these correlations the existence of a causal relationship between temperature and efficiency which may not in fact exist.

Before leaving this discussion of "waking day" experiments, it is of interest to note that the types of performance which have been shown to exhibit diurnal variability include not only tasks of the ·kind already mentioned in the description of the results obtained by workers like Kleitman, Blake, and Pfaff, but also such diverse activities as guessing the time of day (von Eiff *et al.*, 1953); humming (Rubenstein, 1961), where the preferred frequency changes; visual inspection (Colquhoun, 1962); response time to a light signal presented as a secondary task while driving a car (Brown, 1967); and word-association responses (Aarons, 1968). As well as the variation in ability to **estimate** time noted earlier, Thor (1962b) observed a time of day effect on long-range "time perspective" also. In addition to these relatively "high level" functions of the nervous system, diurnal fluctuations have been noted in more "primitive" processes such as the responsiveness of certain sense receptors to stimulation. For example, Jores and Frees (1937) found that the sensitivity of the teeth to painful faradic stimulation gradually rose during the day, reaching its highest point (lowest threshold) at 18.00, and this decrease in pain threshold was confirmed by Kleitman (1963) for the skin. The electrical sensitivity of the eye has also been shown to vary diurnally (Bogoslovsky, 1937; Verkhutina and Efimov, 1947). Thus it is clear that time of (waking) day effects can be detected in functions which involve levels of nervous activity varying from the most simple to the most complex. Since the evidence on these effects is well documented, it is somewhat disquieting to observe the number of psychological studies which are carried out without proper control for them.

## D. *Effects of loss of sleep*

If, as we have argued above, the "waking day" curve of performance efficiency in effect reflects part of a 24-h "sleepiness" cycle, it becomes of interest to know what happens when the **overall level** of sleepiness is increased by depriving subjects of sleep during the time when they would normally have it, and then observing their performance during the following day.

There is good evidence that both the temperature rhythm and the **subjective** sleepiness cycle continue unabated in this situation, not only on the following day, but also throughout a period of sleep-deprivation as long as 98 h (Murray *et al.*, 1958), despite the fact that the **mean level** of temperature falls considerably, and the **mean level** of self-rated "fatigue" rises to a corresponding extent, as the period without sleep is extended. But what about performance? Loveland and Williams (1963) showed that the rate at which simple additions were carried out continued to show a diurnal fluctuation (the speed being greater in the afternoon than in the morning) throughout a period of 3 days of sleep deprivation, and that the variation in efficiency was closely correlated with the (persisting) rhythm of body temperature. The **mean level** of performance, however, declined over the deprivation period along with the mean level of temperature. There was some indication in this study, and in those of Fiorica *et al.* (1968) and Drucker, Cannon and Ware (1969), that the extent of the time of day effect on performance after sleep-loss was **greater** than in control (normal sleep) conditions. Again, in a 12-day "watch-keeping" trial reported by Alluisi and Chiles (1967), 40–44 h of sleep deprivation were introduced at a particular stage of the experiment, and "Diurnal cycling in the performance measures was generally not apparent, **except** during the period of sleep-loss stress" (emphasis added).

All these findings suggest that the effects on efficiency of short-term fluctuations in arousal may be greater when the **overall level** of arousal is low than when it is relatively high. This is in line with the model originally proposed by Freeman (1948), in which the function relating arousal and efficiency is assumed to be non-linear, and, typically, of an "inverted-U" shape (see Fig. 2).

The arousal function plotted in Fig. 2 is of course an idealized one; nevertheless, so long as the actual function conforms in general to the shape indicated, it remains true that, under normal circumstances, the variation in performance resulting from a given change in arousal will be greater when the overall level of arousal is lowered. Only when a prior state of "hyper-arousal" exists, and the level is post-optimal for efficient performance, will this rule break down; such a state would not be expected to occur under the conditions in which experiments on circadian periodicity are usually carried out (cf. Adkins, 1964).

## III. "Round the Clock" Studies

The impetus for research into variations in "true" performance efficiency during the normal "waking day" has come largely from the purely scientific interest of individual investigators, and its application to the practical problems of real-life work in factory and office is perhaps not always easy to appreciate. By contrast, the study of operator variability "round the clock" was prompted primarily by a real and pressing need to know the extent to which human error was likely to affect the running of expensive and potentially dangerous plant and equipment which, for economic or military reasons, it is necessary to operate on a 24-h basis.

Apart from the undoubted rise in the amount of capital invested over the last few years in equipment of this kind, there is another reason which has prompted scientific investigations of human performance under the

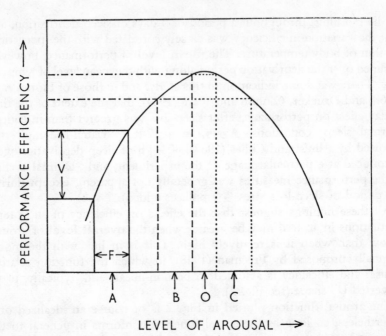

FIG. 2. Postulated relationship between level of arousal and performance efficiency. A given fluctuation ($F$) in arousal level will result in a greater variation ($V$) in performance efficiency when the overall level of arousal is low ($A$) than when it is relatively high ($B$ or $C$) and near the optimal point ($O$) for the particular task.

abnormal living routines necessarily imposed by "round the clock" working. In earlier days the actual number of men employed on duties involving night hours (bakers, seafaring "lookouts", and sentries, for example) constituted only a small fraction of the total working population. It is quite possible that these men were self-selected by an inherent preference for this kind of work (possibly they were good "adapters"); and even if they were not, in most cases the duties they were required to perform were not of the kind in which a simple mistake could involve the loss of anything equivalent to, say, a petrochemical refinery or a nuclear war.

The situation is very different today, where very large numbers of men (and sometimes women) are often required to carry out their work at times when the rest of society is either fast asleep or enjoying itself. Many of these people are **not** self-selected for shift-work; they have to do it whether they like it or not, and, if present trends continue, they, and an increasing number of their fellows, will have to do it even more frequently in the future. When on duty some of these people are engaged in activities which make them directly responsible not only for the efficient operation of extremely expensive equipment, but for the safety of very many human lives as well (early-warning radar operators, nuclear power plant operators, etc.); an increasing pressure is being put on rapid and accurate response to the first signs of danger or of deviation from a norm. Thus night workers must remain as continually alert and vigilant as their daytime counterparts, at a time when their natural tendency would be to sleep. The question then is, can one expect such people to function at the same high level of efficiency as they are assumed to do in the course of an ordinary working day? If the answer to this is "no", then the problem becomes one of either re-designing particular system operations so that an increased risk of human failure can be tolerated (in some cases this would be impossible, in others prohibitively expensive), or attempting to improve the night-time efficiency of the worker himself.

Perhaps not surprisingly, much of the research concerned with human performance under abnormal routines has been sponsored by military or para-military agencies. Thus many of the investigations have been concerned primarily with the problem of optimizing 24-h work-rest schedules for particular circumstances which often have no direct parallel in the industrial field. Nevertheless, the general principles that have emerged from these studies are clearly relevant to the organization of round-the-clock operations in all spheres of activity, whether these be of mainly military significance or not.

A. *Field investigations*

1. *Performance "on the job"*. As in the case of "waking day" studies, we cannot, when assessing the extent to which true efficiency varies round the clock, place too much weight on actual measures of performance recorded in real-life situations, because of the very many uncontrolled variables that are inevitably present. Nevertheless, it is of interest to note that there is general agreement in these "on the job" investigations that efficiency levels are lower in night shifts than at other times. Thus Browne (1949) observed that the delay before switchboard operators answered calls was both greater and more variable during a night shift (23.00–08.00) than during either a day shift (08.00–16.00) or an evening shift (16.00–23.00). Changes within each shift were also observed (see Fig. 3) which, when compared with the circadian temperature rhythm shown in Fig. 1, suggest that the fluctuations in the performance indices used by Browne were quite closely correlated with this rhythm throughout the 24-h period.

Bjerner, Holm, and Swensson (1955) analysed all logging errors made in a Swedish gas works over a period of 20–30 years. During the night

c

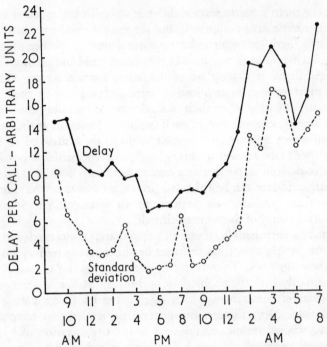

Fig. 3. Delay by switchboard operators in answering calls at different times of day or night. Reproduced with permission from National Institute of Industrial Psychology from Browne (1949).

shift (22.00–06.00) there was a steady increase in the number of errors, which reached a peak at about 03.00, i.e., at about the same time as lowest performance was recorded from the switchboard operators in Browne's study. The morning (06.00–14.00) and afternoon (14.00–22.00) shifts showed no very definite pattern, but there was one interesting aspect of the afternoon shift trend noted in this, and in 2 subsequent studies by the same team; in each case there was a small peak in errors at 15.00. In a review of these investigations Murrell (1965) observed that this peak might be accounted for by the "usual large Swedish mid-day meal" taken before coming on shift. In view of the studies by Blake on the relation of eating to the post-lunch effect which were described earlier, we cannot be certain that the meal itself was responsible for this error peak, but it is nevertheless gratifying to find an industrial parallel to a laboratory-demonstrated effect, whatever its actual cause.

A study of shift-work in a steel mill (Oginski, 1966) showed that, although overall **production** was approximately equal in morning, afternoon/evening, and night shifts, more detailed analyses revealed a superiority of the afternoon/evening shift over the others. Thus reaction-time (recorded in an interpolated test) was fastest on this shift, major accidents were less frequent, and the incidence of shutdowns was lowest. There was no consistent difference between the morning and night shifts in these various measures, but since the morning shift commenced at 06.00 in this factory it could perhaps be described as a "part-night" shift.

Other "on the job" investigations of this type have tended to reach similar conclusions concerning the superiority of certain working periods in 3-shift systems (Menzel, 1962; Ulich, 1964) or in 2-shift (night and day) systems (Wyatt and Marriott, 1953), although the afternoon/evening shift is sometimes found to be socially unpopular, as is the night shift (de la Mare and Walker, 1968).

Gavrilescu *et al.* (1967) were able to demonstrate a change in the within-shift pattern of performance efficiency at an interpolated reaction-time test during the course of a week on night shift in a rubber tyre plant. The decreased decrement in $RT$ they observed was unaccompanied by any corresponding change in either the working or the resting pulse-rate curves. The authors point out that this finding that "higher nervous processes" adapt more quickly to an abnormal routine than "vegetative" ones emphasizes the importance of the central nervous system in physiological adaptation to such routines, and conclude that their results support the view that the physiological mechanisms of the 24-h rhythm of body function are based on the achievement of a "stereotype of time-conditioned reflexes".

2. *Temperature changes.* It is unfortunate that none of the authors referred to above who were able to obtain measures of performance efficiency on various shifts succeeded at the same time in recording the body-temperature rhythms of their subjects. On the other hand, it is equally unfortunate that, in other studies where temperature **was** recorded, there are no corresponding performance data. In these latter studies it has been usual to take temperatures at intervals during a spell of duty on, e.g. a "night" shift, and by plotting the readings over a number of such shifts, to determine the extent to which the subjects were "adapted" to unusual hours of work. The assumption here is, of course, that efficiency levels can be inferred from the temperature curves; however, we have already seen that, over the "waking day" at least, this assumption is only valid in a relatively gross sense. Ideally, therefore, we would prefer to consider only those cases where performance data were actually recorded, but, since these are relatively few in number, we have perforce to include the results of some of these purely "physiological" studies also.

In a series of experiments by Lobban and her colleagues (summarized in Lobban, 1965) conducted in the Arctic regions where, during certain times of the year, environmental time clues from changes in daylight are minimized, it was found that the body temperature rhythm showed, in nearly every subject, rapid and complete adaptation to sleep-activity routines which were "expanded" or "contracted" by varying amounts in relation to solar time. Although this is not quite the same as a **displacement** of routine, Sharp (1961) demonstrated that under similar conditions inversion of the temperature curve in response to a 12-h displacement occurred in 3–4 days in 8 subjects. However, as van Loon (1963) has emphasized, for the solution of practical problems of shift work it is necessary to study shift workers under their "normal" (i.e. local) conditions, where the environmental clues, or *Zeitgebers* (Aschoff, 1960), which appear to synchronize the indigenous circadian rhythms to an exact 24-h periodicity, are, of course, continuously present.

In some of the studies conducted under such normal conditions, inversion of the temperature rhythm in night workers has indeed been found (e.g., Toulouse and Pieron, 1907); in others it has not (e.g., Benedict and Snell, 1902). The conflicting results of early investigations in this area have been reviewed by Teleky (1943), who holds that both the rate and the extent of inversion depends on the amount of bodily activity undertaken during the work spell; and also more recently by Kleitman (1963), and van Loon (1963). The latter author suggests that the essential cause of the contradictory conclusions reached by different observers is the relative infrequency with which temperature readings have been taken; he points out that this inevitably leads to difficulty in recognizing if and when a change in the rhythm has taken place. Kleitman, on the other hand, ascribes failures to demonstrate inversion to the tendency of investigators to expect results in a few days only, in a situation where individual differences in resistance to phase-change (possibly due to age factors) are so great that it is necessary to continue observations for a much longer period. This, of course, would be particularly important where the sample size is restricted for practical reasons; in fact, if Kleitman is correct, whether or not changes in the temperature rhythm are observed will, with the very small number of subjects typically studied, depend very much on the "luck of the draw".

Van Loon himself measured the temperatures of only 3 subjects (van Loon, 1963); however, luck seems to have been on his side, for he was able to demonstrate, if not inversion, at least some clear signs of alterations in the general characteristics of the rhythm when a night shift was being worked. His paper contains several interesting observations, from which we can make the following general statements concerning the difficulties associated with the interpretation of temperature data of this kind.

(i) The essential change in the rhythm after night work is one of **"flattening"** rather than inversion; thus one should not take a **shift in phase** as the primary criterion for deciding whether or not adaptation has occurred (this may account for some of the reported "failures" in the literature).

(ii) Most workers take at least 1, and usually 2 days off each week, during which they return to normal sleeping patterns; this very probably results in a complete reversion to a normal temperature rhythm also. Thus data collected over a period of several weeks from night shift workers and not appearing to show evidence of rhythm change may have been contaminated by reversion due to weekends or other days off.

(iii) It is necessary to distinguish between "short-term" and "long-term" adaptation; the latter may reveal itself as a decrease in the time before the "flattening" appears on each successive period of night working rather than as a more pronounced flattening *per se*.

A physiological study of a kind rather different from those mentioned so far was carried out by Utterback and Ludwig (cited by Kleitman, 1963). They compared 2 "watchkeeping" routines in U.S. Navy submarines in which 3 operators were available to provide continuous monitoring for each particular station by working in turns. In 1 system, each man followed a "4-on, 8-off" schedule (the effects on **performance efficiency** of maintaining this schedule have been studied by other workers, and also by us in our own laboratory, and will be described later). Very little adaptation of the temperature rhythm was observed under this system even after 18 days, the curves being "irregularly bimodal". On the other hand, in an alternative "close" system in which the 8 h of daily duty were compressed into a single 12-h period, distinct "new" 24-h temperature rhythms were established (after varying time intervals) for each of the 3 sections of the crew. Thus, in that section whose watch-standing hours were at night, complete inversion of the temperature rhythm was in fact successfully induced.

A final group of studies which should be mentioned consists of those in which investigations have been made of changes in circadian rhythms in people subjected to multiple time-zone transitions as a result of lengthy journeys in an East–West or West–East direction. There seems to be general agreement that adaptation of physiological rhythms to the new "local" time after such journeys occurs in some 3–5 days (e.g., Gibson, 1905; Burton, 1956; Hauty and Adams, 1965). There is also evidence that indices of performance efficiency tend to follow the physiological trends (Hauty and Adams, 1965; Klein *et al.*, 1970). These results are of great interest in view of the rapidly increasing growth of air travel, and re-demonstrate what the previously mentioned studies in the Arctic, and also those in caves (Kleitman, 1963), bunkers (Aschoff and Wever, 1962), or simulated space capsules (Hauty, 1962) have shown, namely, that the *Zeitgebers* provided by the daily cycle of darkness and light are of primary importance in the entrainment of circadian periodicities.

In summary, past field investigations of "round the clock" operations can be divided into two classes. In the first of these, performance measures were taken from workers on different shifts, but no record was made of body-temperature rhythms; in the second, temperature rhythms were recorded without any accompanying measures of performance. No single investigation in which temperature and performance were **both** observed appears to have been undertaken in "normal" conditions. This is unfortunate, but given the assumption that temperature reflects efficiency during night hours at least to the same extent as it is known to do during the day, we can tentatively conclude

  (i) that performance on night shifts is generally worse than on day shifts,

  (ii) that, when the sleep–waking cycle is altered, little change occurs in the temperature rhythm for several days, following which the curve tends to "flatten", and thus

  (iii) that efficiency trends during a number of consecutive nightshifts should be characterized by a gradual change from pronounced within-shift decrement to relatively constant performance throughout the duty spell. Validations of these conclusions are considered in the discussion of the pertinent laboratory experiments which now follows.

### B. *Laboratory experiments*

When subjects are confined in a laboratory it is much easier to take concurrent measures of both performance efficiency and physiological state than it is in field investigations. Also, as was said earlier, the increased degree of control that it is possible to exercise over the situation means that the performance scores obtained are more likely to reflect "true" efficiency than those which are normally available on the shop floor. In addition, the physiological observations will almost certainly be more reliable. The present section is concerned solely with experiments conducted under these controlled conditions.

1. *Research findings*. The first of these investigations that we should consider is one carried out by Kleitman and Jackson (1950). They recorded changes in performance efficiency and in body temperature in 9 young male subjects who were required to follow certain watchkeeping duty-schedules which were either already in use in certain ships of the U.S.

Navy, or were being evaluated for such use. In one of these schedules the subjects had 24 h of duty in each of 5 successive 96-h periods, but the work-spells were distributed in a rotating manner over 5 4-h shifts between 20.00 and 16.00 the next day, and 2 2-h intervals between 16.00 and 20.00, when the watches were "dogged". Sleep was taken at different hours on successive days, in the longest and most convenient intervals between work periods. Compilation of the results obtained in this rapidly rotating kind of system over 1 or more complete shift cycles enables a picture to be obtained of "round the clock" variation in the levels of all physiological and psychological measurements taken. Kleitman and Jackson's study of this system provided a test of whether the relationship between temperature and performance observed by Kleitman in his "waking day" studies described earlier in this chapter was maintained during the hours when the subjects would normally have been sleeping.

Three tests of performance were used to measure efficiency. These were (i) flying a "Link Trainer", (ii) complex choice reaction-time, and (iii) colour-naming. The last test showed the closest association with body temperature, the mean time taken to name 600 colours varying inversely with mean temperature as the latter fluctuated throughout the 24-h period (see Fig. 4). Note that the temperature continued to follow its normal rhythm in this shift system.

Performance at the Link Trainer and the complex choice reaction–time tests was quite well related to temperature during "day" hours, but not at night, where in the former test unexpectedly high efficiency scores were obtained, and in the latter the results were "erratic".

Kleitman and Jackson also examined several alternative watch-keeping systems in their investigation. All of these systems differed from the one already described in a major respect—in each case the hours of work remained **fixed in time** throughout the observation period. This kind of system, which we shall refer to as "stabilized", requires a change in the sleep-waking cycle when any of the work periods fall during "night" hours. Kleitman and Jackson found, in comparing various versions of this system, that the less the deviation¦ from the usual routine, the better was the adjustment of the body temperature curves to the new cycle. They also concluded that "in a general way, the higher the body temperature, the better was the performance", but they did not, apparently, obtain the same degree of closeness of association between temperature and efficiency in these stabilized systems as they did in the rotating system. We shall be looking more closely at stabilized systems later, so we shall leave discussion of this suggested difference between rotating and stabilized systems for the moment, noting meanwhile the important fact that Kleitman and Jackson were able to demonstrate a relatively high degree of co-variation between

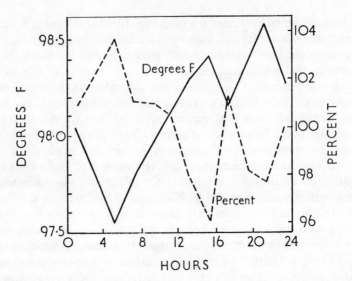

FIG. 4. Variations in group mean oral temperature taken at the hours of test administration shown on the abscissa, and concomitant variations in group mean time required to name 600 colours, the latter expressed as a percentage of the total group mean score for the period. Reproduced with permission from Kleitman and Jackson (1950).

temperature and at least one aspect of performance, not just during the "waking" day, but **right round the clock.** In so far as temperature reflects the "sleepiness" of the individual this result is of course exactly what would be predicted on the hypothesis that performance efficiency at any particular point in the 24-h period is directly related to the level of arousal at the time.

An 8-year programme of research on the effects on efficiency of various work–rest schedules during confinement in a simulated aerospace vehicle crew compartment (Chiles, Alluisi, and Adams, 1968) provided, amongst other things, some interesting data on diurnal cycling of performance over prolonged periods of observation. In summarizing this aspect of the results, Alluisi and Chiles (1967) concluded that "the diurnal rhythm which is evidenced in physiological measures may also be evidenced in the performance depending on the information given to, and the motivation of, the subjects, and depending also on the total work load; even where motivation is sufficiently high, the cycling may be demonstrated when the operator is overloaded". They were led to this conclusion as a result of their observations

(a) that performance in certain **task-situations** exhibited diurnal variation, whereas in others it did not,

(b) that the efficiency of certain **groups of subjects** fluctuated in a circadian manner, while that of others remained constant, and

(c) that, during a prolonged "flight", cyclic effects were more likely to appear at **particular stages** than at others.

They also noted that loss of sleep tended to produce an occurrence of diurnal periodicity in performance where this was not previously apparent.

In considering all these various factors, Alluisi and Chiles put forward the concept of "performance stress", which is postulated to exist when the situation presents the operator with a work-load that is more than he can handle. In complex tasks this state of stress will, it is argued, often be present during the learning period, and it is at this time that diurnal cycling of performance may appear, only to vanish at a later stage of practice, when the task has been mastered. At the same time, the inherent liability of even simple tasks to exhibit circadian periodicity may, it is thought, be overcome by a sufficiently high level of motivation in the subject. However, limits to the effectiveness of such motivation are acknowledged, and diurnal variation may thereforere appear when the operator is either overloaded by the task requirements, or subjected to stress, e.g., by deprivation of sleep.

What Alluisi and Chiles are saying here ties up very well with the "arousal" theory of performance, and the way in which diurnal fluctuations in arousal may or may not be reflected in variations in efficiency, depending on which part of the "inverted-U" curve the subject is operating (cf. Fig. 2). If we suppose that the level of arousal is increased by task-complexity and motivation, but decreased by loss of sleep (note that we do not regard sleep deprivation **itself** as a "stress" in Alluisi and Chiles' sense), then it can readily be seen that certain combinations of these factors (e.g., simple task, reasonable motivation) will produce a level of arousal around the optimum, when little or no diurnal variation will be evident. Other combinations (e.g., complex unpractised task, very high motivation) may, on the other hand, result in a state of **hyper-arousal**, when fluctuations due to time of day will be relatively pronounced. Circadian changes will also be apparent at the opposite (lower) end of the arousal curve; for example, when loss of sleep is combined with a simple task in the absence of specific incentives.

The main trials in the series of experiments by Chiles and his colleagues required the subjects to work either on a "4-on, 4-off" or on a "4-on, 2-off" schedule. The shift systems were thus "stabilized", but because of the high work–rest ratio and the relatively short work periods, several discrete sets of both physiological and performance measures were obtained from each man during each 24-h period. Since the individual subjects' schedules were staggered, it was possible to compute "round the

clock" curves of these measures, comparable to those obtained from
"rotating" systems, by averaging over days and subjects. The mean
rhythms of body temperature thus derived exhibited the usual circadian
form, but there was some evidence that towards the end of the longest
trial (which continued for 30 days) there was a slight reduction in the
amplitude of the rhythm. There was also an apparent shift of some 2–4 h
in the **phase** of the rhythm as this particular trial progressed (peak temperature tending to occur later), but this phase-change was not subjected to
statistical test. No evidence of the development of polycyclic curves was
found.

Daily variations in several of the performance measures (which included
tests of vigilance, arithmetic computation, problem-solving and pattern
recognition) were observed to parallel, at least approximately, the physiological periodicity. The peak within-day performance scores occured at
about the same time as the peak temperature reading during the early
period of a given study; however, later in the study the performance peak
appeared to lag the physiological one by 2–4 h. Again, no statistical test
was made of this lag, so, as Chiles and his co-authors themselves say "the
implications of this finding (if it is real) are interesting but they can be
accepted only tentatively at present" (Chiles, Alluisi and Adams, 1968).
Of possible relevance here are the results of an investigation by Thompson
(1967) of rhythms in several physiological and psychological variables
recorded at 4-h intervals. Although in this case the observation period
was only 2 days a similar lag was noted in 2 of the 5 subjects tested. However, this lag was "inclined to be variable and phase locking (did) occur".
Twenty-four hour periodicity was observed in auditory threshold, reaction-
time, and critical fusion frequency (CFF) in this experiment.

Several other studies have been reported in which various performance
measures have been recorded at times both inside and outside the usual
working day. In most of the experiments some at least of the tests were
made at times when the subjects would otherwise have been asleep; thus
it is not altogether unreasonable to classify these studies as "round the
clock" for present purposes. Although few investigations have approached
in scale the magnitude of the trials conducted by Chiles and his colleagues,
the results obtained confirm, in general, the existence of 24-h rhythmicity
in a wide range of activities such as, for example, simple letter cancellation (Fort, 1968); visual *RT*, and also time estimation (Adkins, 1964),
and sensorimotor co-ordination (Klein, Wegmann, and Bruner, 1968).
Most authors seem to agree that the "low point" in alertness and efficiency
occurs somewhere during what would normally be the subject's sleep
hours, and typically at about 04.00, which tends to correspond with the
time of minimum activation as measured by physiological indices.

It should be noted here that although in some experiments of the "round the clock" type subjects are awakened from sleep for testing in the "small hours", the results obtained show no obvious differences from other experiments in which they remain awake beforehand. This is important, since it demonstrates that the basic cause of poor performance recorded at these times cannot be the fact that normal sleep has been **interrupted**. It is nevertheless highly probable that the sleep of those subjects who are awakened will have been shorter than usual, and thus in both situations a state of **sleep deprivation** will exist at the time of testing in these "night" sessions. In order to control for the confounding effects of this sleep-loss on performance it would be necessary to have the subjects follow a routine which involved their going to bed several hours earlier than usual. As we shall see later, the difficulties of arranging for this are considerable when, as is the case with most investigations in this area, the subjects have to live, outside their "working" hours, in a social environment where other people are following a normal routine; and even if these difficulties are overcome (for example, by the use of specially constructed "living units" in which the subjects are completely isolated from the outside world for the duration of the experiment) there is no guarantee that actual **sleep** will start at the prescribed time.

2. *Techniques of analysis.* One of the points most frequently commented on by authors is the substantial extent of inter-subject variability that is encountered both in the amplitude and in the phasing of the performance rhythms that are found to exist. We have already suggested the way in which differences in motivational level could affect the magnitude of the 24-h variation in arousal that would be expected, and this factor might account for at least some of the observed differences in the **amplitude** of performance rhythms. As regards the differences in **phasing**, Kleitman (1963) was among the first to draw attention to the existence of "morning" and "evening" types, and we shall see in the following chapter that it is now possible to distinguish these 2 classes of people not only in physiological and performance terms, but in measurable personality characteristics as well.

A question that is related to the assessment of the extent and nature of individual differences in circadian performance rhythms is the **method of analysis** that is used to establish the existence of periodicities in the data in the first place. This is, of course, a basic problem, not only in the present context, but in the whole field of biological rhythm research. Although in recent years several sophisticated mathematical techniques have been evolved for the determination of periodicities in time-structured observations (see Chapter 1), these techniques have been applied only rarely to performance data of the kind we have been considering. This is partly because observations are seldom made at a frequency which would appear to justify treatment of the results in this kind of way; it is also partly because performance scores are inherently more "noisy" than physiological measurements, since the difficulties of holding constant certain factors which may affect efficiency (e.g., motivational level) are very great, particularly when the period of observation is at all prolonged. However, it must be admitted that simple lack of knowledge on the part of the experimenter has, in many cases, probably been the main reason why the analytical methods employed have been relatively crude. The investigator working in the field of periodicity in performance today is in many ways in a position which resembles that of his physiological counterpart some time ago; the data are typically plotted over 1 or more cycles and visual inspection (sometimes supplemented by "conventional" statistical tests, e.g., of the significance of the mean difference between selected time points) is made the basis of any conclusion drawn. Although it is certainly true that, as Halberg (1965) says "important information has been derived by students of rhythms relying heavily upon the gross inspection of biologic data plotted as a function of time", it nevertheless appears likely that the rate at which progress in the understanding of the nature of rhythmicities in performance efficiency takes place will depend to a large measure on the extent to which the more advanced analytical techniques now available are applied to the results of experiments.

The above view is evidently shared by Frazier, Rummel and Lipscomb (1968), who, after reviewing relevant studies on the performance of (in this case) vigilance tasks, and concluding that these suggest that such performance "can show circadian variations", go on to remark that "in the absence of results from analyses which yield information specifically about period, phase, and amplitude relationships there is very little else (about this variation) that can be concluded at this point". Frazier and his colleagues themselves used least-squares spectral analyses to evaluate the results of their own experiment. Their reasons for choosing this particular technique, rather than one of a number of alternatives, are summarized by them as follows: "Data can be analysed on an individual basis and missing data can be tolerated which is beneficial for use in non-laboratory situations. Slow fluctuations in rhythmicity do not preclude the use of the analysis since large quantities of data are not required, in contrast to the quantities of data required by alternative analytic techniques for discrimination of periodicity. With alternative techniques a premium is placed on the assumption of stationarity. The use of an incremental block analysis also can provide an indication of how a given day's events might alter rhythmicity, which would be very useful in some experimental situations. Finally, the analytic power of this technique is relatively substantial because it reveals not only whether circadian rhythmicity is present as a substantial source of performance variation, but also how these periods can be described in terms of amplitudes of stable and varying components, and phase relations".

This is an impressive testimonial for the usefulness of the least-squares spectral analysis technique; Frazier and his co-workers claim that, in their case, its application revealed "circadian rhythmicity . . . in every measure employed". They also found that "individual circadian periods showed clear nonstationarity as time progressed, with periods ranging considerably above and below 24 h". However, one cannot help wondering whether the conclusions drawn by Frazier *et al.* (1968) are really justified on the basis of the data that they collected. In the nature of the experiment, observations of performance levels were not undertaken during the subjects' sleeping hours (which remained normal throughout the experiment), and irregular sampling of this kind is not exactly in the best interests of spectral analysis. Furthermore, even during the waking period only 4 tests were made. It is therefore not clear what significance can be attached to statements such that "periods ranged considerably above and below 24 h". Presumably it can be accepted that some degree of non-stationarity was indeed present, but the meaning of the occurrence of, for example, a 30 h periodicity in a subject going to sleep at the same time each night is not immediately obvious, and the layman might thus be forgiven for being slightly sceptical of this kind of approach.

The foregoing account of laboratory experiments on "round the clock" performance is not meant to be exhaustive; for further information the reader is referred to an excellent review by Trumbull (1966), where several other pertinent reports (some of which have not been published in the usual scientific journals) are listed. However, from what has been described it would appear fairly clear that many aspects of performance do in fact show a pattern of fluctuation over the 24-h period which corresponds, at least approximately, with the underlying sleep–wakefulness cycle under "normal" conditions, i.e., when no attempt has been made to change the course of this cycle by a systematic alteration of the hours of work and rest. We now turn to those studies in which the experimenter has set out to do just this, in order to simulate in the laboratory the kind of routine followed by shift workers in real life.

## IV. Simulated Shiftwork

Apart from the study of "close" Naval watchkeeping systems by Kleitman and Jackson mentioned earlier, the present writer is aware of only one other published investigation in which a laboratory experiment has attempted to reproduce the kind of living routine that might be followed by industrial night-shift workers in real life, in order to observe the effects of the resultant alteration in the sleep–rest cycle on performance efficiency. This investigation was conducted by Wittersheim, Grivel and Metz (1958), unfortunately on only 2 subjects. A multiple-choice reaction task of 25 min duration was performed on 5 successive days at 09.00, 11.00, 15.00, and 17.00. Normal sleep was taken during the intervening nights. A week later the series was repeated, but this time the tests were at 23.00, 01.00, 04.00 and 06.00, the subjects now sleeping during the intervening **days.**

A decline in efficiency during the night was evident at the start of the "night" series in both subjects, but whereas in one the extent of this decline diminished only slowly throughout the observation period, in the other it decreased in 2 of the measures (speed and errors) much more rapidly. Since it was also noticed that the latter subject slept better during the day than the former, it was concluded that the difference between them was due to a difference in their ability to adapt to the inverted sleep–rest cycle. No physiological data are given in Wittersheim's report, so it cannot be ascertained whether this difference was reflected in the temperature rhythms of the subjects. Kleitman and Jackson (1950) found some diminution in the amplitude of the temperature rhythm in their "night work" (24.00 to 12.00) trial, in which the subjects continued on their abnormal living routine for over twice as long as the 2 subjects in Wittersheim's study; but since within-watch trends in test scores are not detailed in Kleitman and Jackson's report, in this case it cannot be determined whether adaptation occurred in **performance.**

In the present section we shall be describing our own experiments in which both temperature **and** performance were recorded at regular intervals in a series of prolonged trials. In some of these trials the work–rest routine paralleled very closely that followed by shift workers in industry, although the investigation (which was carried out under the aegis of the British Royal Naval Personnel Research Committee) was primarily concerned with the evolution of optimum "watch keeping" systems for personnel engaged on maritime operations. The results of these experiments have been described fully elsewhere (Colquhoun, Blake and Edwards, 1968a, 1968b, 1969); the general procedure followed was based on methods developed in a pioneering study by Wilkinson and Edwards (1968).

A. *Experiment* 1: *"Rotating" versus "stabilized" split-shift systems*

1. *Method.* During routine periods of operation, Naval personnel are frequently divided into 3 teams, or "watches". It is customary to rotate

duty periods among these teams in a manner which results in fair alloca-
tion of unpopular work times to all concerned within a reasonable time
span. Although there are minor variations in the way this is done, the day
is typically divided into 6 4-h periods starting from 24.00, and each team
is employed once on each of these "shifts" during a 72-h cycle. The
inevitable results of the operation of this system is that the times at which
duty spells and rest periods occur are never the same for members of a
particular team on successive days. This kind of system is very similar to
one that Kleitman and Jackson (1950) studied in their investigation
described earlier.

The first experiment compared the performance of men working at
mental tasks under this "rotating" system with that of other men working
for shifts of similar length under an alternative, "stabilized" system (also
examined by Kleitman and Jackson) in which the hours of work and rest
were the same each day. Twenty-eight young Naval ratings volunteered
to take part in this experiment. Twelve subjects followed a routine in
which the cycle of 4-h shifts was as follows:

| Day 1 | Day 2 | Day 3 |
|-------|-------|-------|
| 24.00–04.00 | 04.00–08.00 | 12.30–16.30 |
| 08.00–12.00 | 16.00–20.00 | |
| 20.00–24.00 | | |

The experiment started at 08.00 on Day 1, after a normal night's
sleep. The Day 3 shift commenced at 12.30 rather than 12.00 in order to
facilitate local messing arrangements, which provided for a substantial
meal at 11.45.

Sixteen subjects were assigned to the stabilized system. The actual
shifts worked were 24.00–04.00 and 12.30–16.30, commencing on Day 1
with the latter shift, after a normal night's sleep. The choice of these
particular shifts was motivated primarily by 2 considerations:

(a) they should coincide with shifts worked in the rotating system, in
order to facilitate intersystem comparisons, and

(b) they should include both a "day-time" and a "night-time" period
of work, as a test of adaptation to the latter.

Clearly it would be possible to evolve stabilized systems other than
the one implied by the shifts chosen (which is in fact a "4-on, 8-off"
system) without resorting to duty spells of greater duration than 4 h.
However, since this system is the simplest one, it was felt that it would
provide a fair test of the advantages of stabilization, while avoiding
practical difficulties which might arise in the administration of more

complicated systems. Of the 3 pairs of 4-h shifts within a 4-on, 8-off system geared to 24.00, the pair selected for study is the only one in which the periods of work coincide unambiguously with day-time and night-time hours.

Apart from the work–rest schedule, the routine under both rotating and stabilized systems was identical. Subjects, who were always tested in small groups of between 4 and 6 individuals, worked up to 23 shifts over a period of 12 consecutive days. Each 4-h shift was divided into 2 work periods of equal length by a 10-min tea break. Oral temperatures were recorded by clinical thermometer (3 min insertion) at the beginning and end of each shift, and also at the start of the tea break. In the rotating system, sleep was allowed in any off-duty period; in the stabilized system, sleep was scheduled from 04.30 to 11.30 (no significant difference was observed between the systems in average **total** hours slept during the trials). A certain amount of additional light duty was required of the subjects in some of their off-duty periods.

During the first work period, subjects performed an auditory discrimination task (the "Vigilance" test) resembling certain sonar operations. Scores on this test were

(i) the number of signals detected (expressed as a percentage of the number presented),

(ii) mean response latency to the detected signals, and

(iii) the number of "false" reports made.

During the second work period subjects carried out simple additions (the "Calculations" test). An *ad lib.* supply of printed pages, each containing 125 columns of 5 2-digit numbers, was made available to the subject, who entered the sum of each column at its foot. Subjects were instructed to work as fast as possible without stopping, and without making mistakes. The 2 scores on this test were

(i) the number of sums attempted (i.e., rate of work, or output) and

(ii) the percentage of incorrect answers given (i.e., error).

Both tests were performed in isolating booths measuring approximately 3 ft ×3 ft ×6 ft. The mean of the oral temperatures recorded at the beginning and at the end of each work period was taken as the average level during the intervening test(s) in this, and in all subsequent experiments.

## 2. Results

(a) The rotating system. All scores showed considerable changes from shift to shift as the trial progressed. The fluctuations in efficiency appeared, in the case of certain scores, to be related to concurrent alterations in body temperature resulting from its circadian rhythm, which persisted with constant amplitude and phase throughout the trial. All parts of this rhythm were sampled during each 72-h cycle; the range of temperature observed was therefore maximal (*circa* 1·0° F). In general, scores were lowest at low temperatures (e.g. between 02.00 and 08.00) and highest at high temperatures (e.g. between 16.00 and 22.00); this variation was superimposed on an underlying "practice" curve, as is illustrated by the output score in the Calculations test, shown in Fig. 5. In this figure the close correspondence between changes in average rate of work and changes in average body temperature is clearly evident.

As a test of the consistency of the relationship exhibited in Fig. 5, each subject's results were examined separately using a multivariate regression analysis, in which the regression of output score on temperature was computed after allowing for linear, quadratic and cubic practice effects in the former.

The regression coefficient was found to be positive in sign in 10 of the 12 subjects, and statistically significant at an acceptable level of confidence in 5 of these 10. The coefficient computed from the mean values of output and temperature plotted in Fig. 5 was itself highly significant ($p < 0.001$). Thus, although there was a considerable amount of variation between subjects, it can be concluded that there was, in general, a positive relationship between body temperature and this particular performance score.

A simplified picture of this relationship for the group as a whole was obtained by computing average values of output and temperature for each of the 6 shifts over the 3 complete cycles represented by Days 2–10 of the trial. These averages are shown in Fig. 6, together with the corresponding error scores. Also plotted in Fig. 6 are the similarly computed performance scores and temperatures from the Vigilance test. (Note: since the "practice" curves for the 5 measures of performance provided by the 2 tests were all essentially linear for the group as a whole over Days 2–10, a simple correction has been made to the obtained means to adjust for bias due to the order in which the 6 shifts were worked.)

Apart from 1 reversal during the morning period there was a 1-to-1 correspondence between changes in Calculations output and changes in temperature, the overall range of variation in performance being about 8%. By contrast, it will be seen that fluctuations in the error score in this test were very small, and did not appear to be related to temperature changes.

In the Vigilance test there was a 1-to-1 correspondence between changes in detection rate and changes in temperature, the overall variation in performance being about 13%. Fluctuations in the false report score were small, and apparently unrelated to temperature changes. On the other hand, the mean time to respond to signals did show a distinct rhythm,

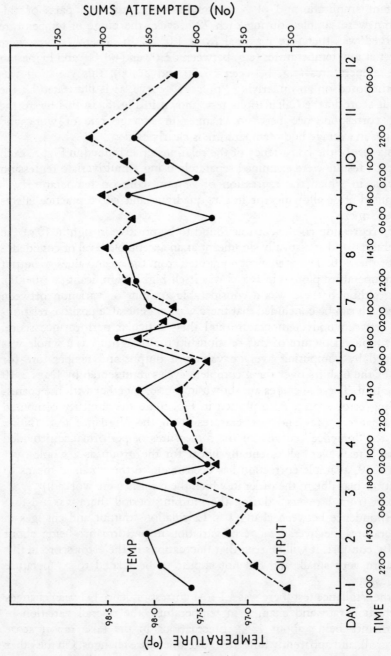

Fig. 5. The rotating system: mean output score in the Calculations test, and mean body temperature, in the second work period of each successive shift. Reproduced with permission from Colquhoun *et al.* (1968a).

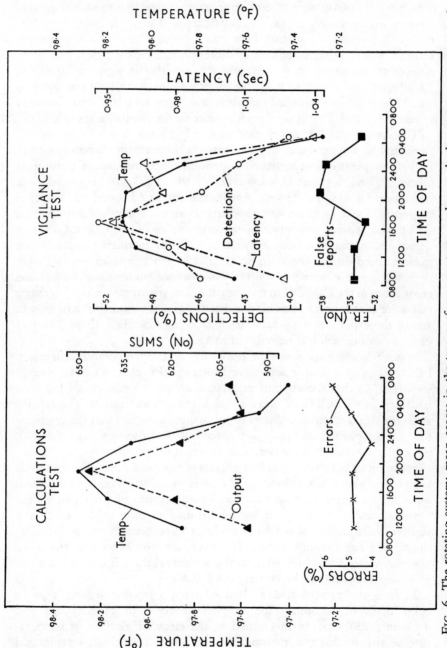

FIG. 6. The rotating system: mean scores in the two performance tests, and mean body temperature, averaged over Days 2–10 for each of the 6 different shifts. Redrawn from Figs 2, 3 and 4 in Colquhoun *et al.* (1968a).

which was in phase with that of temperature (note that the scale for this score in Fig. 6 is inverted, i.e., latency **decreased** with increasing temperature). The range of variation here was approximately 8%. Multivariate analyses showed that in the case of detection rate the coefficient of regression of performance on temperature was positive in sign for 10 of the 12 subjects, and in that of response latency negative in all but one individual. Thus it can be concluded that there was a general relationship between temperature and 2 aspects of performance in the Vigilance test also.

The fact that false report rate in the Vigilance test was apparently unrelated to temperature suggests that the increase in detection rate at higher temperatures represented a genuine improvement in perceptual efficiency, and not merely a lowering of the decision criterion adopted for reporting a signal (Swets, 1964); the increased speed of detection responses could be taken as supportive evidence for this interpretation. Again, the relative constancy of the error score in the Calculations test demonstrates that the subjects were not simply sacrificing accuracy for greater speed during the shifts worked at higher temperatures, but were definitely more efficient at the task as a whole (note: since neither the error score in the Calculation test, nor the false report rate in the Vigilance test were ever found to exhibit consistent signs of circadian fluctuation in any of the experiments in the series now being described, these 2 indices of performance will not be referred to again).

It is of interest to note that the shape of the temperature curves in Fig. 6 suggests a simple sine-wave function. However, we have already seen (Fig. 1) that the rhythm in a large sample of subjects of the type used here is not well represented by such a simple expression. The explanation for this discrepancy almost certainly lies in the fact that the temperature means plotted in Fig. 6 are spaced at 4-h intervals over the 24-h period. It can readily be appreciated that had the readings for the "24-h" curve in Fig. 1 been separated by intervals of comparable length, the resulting picture would correspond quite closely to a sine-wave. It would seem quite likely, therefore, that previous interpretations of the shape of the circadian temperature rhythm have been inadvertently influenced by the relative infrequency with which readings have been taken, particularly during the late evening period. However, the possibility that the more complex wave-form obtained from the subjects at this laboratory is unique to young males cannot be entirely ruled out.

In sum, the present results obtained from a rapidly rotating type of split-shift system confirm the original observations of Kleitman and Jackson (1950) that, in this situation, the circadian rhythm of temperature persists unchanged in amplitude and phase throughout a prolonged experimental period, and that efficiency at mental tasks is quite closely related to this rhythm. "Round the clock" covariation of performance and temperature would thus appear to be a well established phenomenon in the laboratory as well as in the field.

(b) The stabilized system. In order to examine the extent of adaptation of the body-temperature rhythm to the new sleep–waking schedule imposed by the stabilized system, temperatures were recorded over the greater part of the 24-h period on 1 day before the trial began (with normal sleeping hours), and again on the final day of the experiment. The 2 sets of mean readings obtained are shown in Fig. 7, from which it will be seen that the onset of the decline in temperature previously occurring at about 21.00 had, on the final day, moved to about 02.00, i.e., by about 5 h, which was the length of time by which sleeping had been delayed. Thus it appeared that a shift in the **phase** of the temperature rhythm had taken place; since readings were not obtained during the sleep period on the final day it cannot be ascertained whether the **amplitude** of the rhythm altered or not.

As a result of the change in its rhythm the temperature during the "night" shift was now higher than the temperature during the "day" shift, reversing the pattern existing at the start of the trial when the normal circadian fluctuation was occurring. This reversal was more striking in the first work period than in the second, since during the latter period temperature in the night shift continued to fall throughout the trial, in "anticipation" of approaching sleep (cf. Fig. 1).

The manner in which this "adaptation" of the temperature rhythm took place during the 12-day observation period is illustrated in Fig. 8, in which 3-day moving averages (Guilford, 1956, p. 47) of the temperature readings for the first work period of the night and day shifts are shown, together with similarly computed averages of the Vigilance test scores.* All values in Fig. 8 are based on the data from 12 subjects only, since frequent apparatus failure invalidated the results for 1 group of 4 subjects.

The temperature graph suggests that the phase shift of the rhythm commenced on the sixth day and was effectively completed by the eighth. Thus the change was apparently relatively rapid once it was initiated. It is interesting to observe, however, that although there was no evidence of a progressive alteration in the **difference** between "night" and "day"

---

* The performance scores in Fig. 8 (and in all subsequent illustrations in this chapter) are expressed as differences from the daily means. This is convenient for purposes of graphical presentation, since it removes the day-to-day practice effects from the data. Although this procedure does not, of course, remove the practice effect from the **within-day** trends, adjustment of the scores for this would have been not only difficult but also essentially unnecessary, since careful inspection of the original data showed that in all cases the effects of any such adjustment would be negligible.

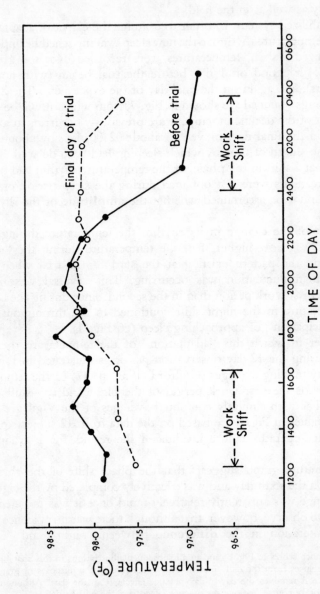

FIG. 7. The stabilized system: mean body-temperature curves before, and on the final day of the trial. Reproduced with permission from Colquhoun et al. (1968a).

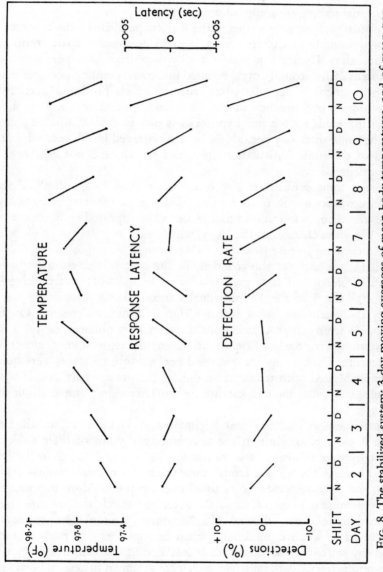

FIG. 8. The stabilized system: 3-day moving averages of mean body temperatures, and of mean performance scores in the Vigilance test, for the night (N) and day (D) shifts.

temperatures over the first 5 days, the **absolute levels** of both readings rose steadily during this period; this rise may represent the first stage of a 2-stage process involved in the adjustment of the temperature rhythm to an abnormal waking–sleeping routine.

Vigilance performance scores in the 2 shifts paralleled the temperature readings throughout the trial with remarkable fidelity, apart from the first three days. During this period it is hypothesized that a persistence of the normal 24-h "arousal" cycle resulted in a corresponding persistence of the "post-lunch" effect (described earlier), which produced a relative depression of performance in the first work period of the day shift. Supporting evidence for this hypothesis is provided by the finding that, over the same three-day period, the scores observed in the **second** work period of this shift (Calculations test, not shown) did **not** exhibit this relative depression.

There is some evidence in Fig. 8 that the rate of "adaptation" of performance was somewhat faster for detection rate than for response latency, in that the changes in trend in the former score appeared to anticipate by about 1 day the changes in the latter, which were precisely in phase with those in the temperature trend from Day 4 onwards.

Multivariate analyses showed that, in the case of detection rate, the coefficient of regression of performance on temperature was positive in sign for all but 1 of the 12 individual subjects. In the case of response latency the coefficient had a negative sign in 8 subjects' results. Of the 16 subjects from whom Calculations test data were obtained only 1 gave a negative sign for the coefficient representing the regression of output on temperature. Thus it appears that the 3 performance measures were quite closely associated with temperature during the greater part of this trial, despite the fact that the temperature rhythm altered in phase at about the halfway mark.

It is somewhat puzzling that Kleitman and Jackson (1950) failed to obtain a clear phase-shift in the temperature rhythm of their subjects when they were following a work routine which was identical to the one studied here. Although the temperature minimum moved from 04.00 to 08.00, the peak remained at its usual evening point. More importantly from the present point of view, the actual on-watch temperature itself remained higher during the day shift than during the night shift **throughout the trial.** A correspondingly persistent superiority of **performance** in the day shift was also observed. It is difficult to explain this difference in the results from 2 experiments which were so similar in design; however, 1 possible factor which might have influenced the outcomes is an apparent difference in the sleeping patterns adopted by the subjects. Whereas in the present experiment all subjects had a single long sleep between 04.30

and 11.30, Kleitman and Jackson state "in the distribution of the hours of sleep, a fairly solid area could be seen between 4 and 8, and a much lighter one in the afternoon and evening". This suggests that the subjects in their investigation were tending to divide their sleep into 2 separate periods. Since a single long sleep period is clearly an integral part of the **normal** 24 h work–rest routine, failure to obtain it in the context of an **abnormal** routine like the present one may well prevent physiological adaptation to the new cycle from taking place.

B. *Experiment 2: Eight-hour stabilized systems*

1. *Method.* It was clear from the results of Experiment 1 that the advantages of stabilization of working hours in a 3-team system would be lost if the period of duty was split into 2 4-h shifts separated by 8 h, since, for each team, 1 of the 2 shifts would necessarily occur at the beginning, and the other at the end of the "physiological" day. They would thus suffer from a near-maximal temperature differential, which in turn would be expected to be reflected in substantial inter-shift variability. The second experiment therefore investigated the effect of reducing the rest period between 2 4-h shifts to a point at which the duty-spell became, in effect, a single 8-h shift with meal break. It was hoped to determine whether stabilized shifts of this type produced, by their very length, effects on performance from "fatigue" which were of sufficient magnitude to offset the theoretical advantage they provide of allowing duty spells to be arranged in such a way as to avoid starting work just after the sleep period, or stopping work shortly before (in non-military situations, of course, it might in practice be very difficult to make such arrangements where shifts were timed to change during "night" hours).

Three 8-h shifts were studied: a "night" shift (22.00–06.00), a "morning" shift (04.00–12.00), and a "day" shift (08.00–16.00). These particular shifts do not represent a single **complete system** of 8-h shift working, but were chosen for investigation because they cover 3 main areas of interest:

(a) night-time working and adaptation to this (night shift),

(b) effects of very early rising, i.e. of phase-**advancing** the work–rest cycle (morning shift) and

(c) correlation of performance and temperature in "control" conditions (day shift).

On the basis of the results obtained in Experiment 1 it was expected that performance levels in the night shift would, initially, fall during the duty-spell as temperature declined according to its natural rhythm, but that, if and when the temperature rhythm altered in response to the new work–rest cycle, a corresponding change in efficiency trends would ensue.

The morning shift was included to investigate whether, despite the anticipated continued **relative** depression of temperature at the start of the shift (consequent on the impracticality, in the experimental circumstances obtaining, of arranging sleep-hours to avoid close juxtaposition of waking up and starting work), some degree of alteration in the temperature rhythm would nevertheless occur on repeated exposure to this phase-advanced work–rest cycle (one which it might in real life be necessary to employ in certain circumstances), and whether, in this case, the alteration was accompanied by corresponding changes in performance.

It was intended to assess the extent, if any, to which performance was affected by "fatigue" effects due to the increased length of the duty-spell by observing, in the day shift, whether the within-shift trends in efficiency differed to any marked extent from those predicted on the basis of the previously demonstrated correlation of efficiency and temperature within the normal rhythm. The day shift performance trends would then act as a base against which the results obtained from the night and morning shifts could be evaluated.

Thirty-one naval ratings volunteered for these trials, which, as in the first experiment, extended over a period of 12 consecutive days. Eleven subjects worked the day shift, the routine for which was as follows:

| Time | Activity |
|------|----------|
| 08.00–09.40 | Work-period 1 |
| 09.40–09.50 | Tea break |
| 09.50–11.30 | Work-period 2 |
| 11.30–12.30 | Meal break |
| 12.30–14.10 | Work-period 3 |
| 14.10–14.20 | Tea break |
| 14.20–16.00 | Work-period 4 |

Sleep time for these subjects was from 23.00 to 06.30, and no additional duties were required of the subjects outside their working hours. Body temperature was recorded at 08.00, 09.40, 11.30, 12.30, 14.10 and 16.00.

The within-shift routine for the night shift and the morning shift was identical with that for the day shift apart from the time of commencement. Ten subjects worked the night shift, sleeping from 08.00 to 15.30, i.e., with a temporal displacement of the sleep period of some 9 h. Ten subjects worked the morning shift. The "official" sleep time for these subjects was from 19.30 to 03.00, but in practice, due to pressures arising from the social environment in which the subjects lived outside the laboratory, sleep rarely commenced before about 22.00. Thus the displacement of the sleep period in this group was about $3\frac{1}{2}$ h if measured from the time of awakening, but only about 1 h if measured from the time of going to sleep; the **amount** of sleep obtained was, also, obviously curtailed.

Each of the 4 work-periods of the shift was divided into 2 consecutive sessions of 50 min duration. During the first session subjects performed the Vigilance test, and during the second session they worked continuously at the Calculations test.

## 2. *Results*

(a) Day shift. In order to facilitate comparison with the results from Experiment 1, 3-day moving averages of performance scores and temperatures were again determined. These averages are shown in Fig. 9, in which the top graph represents temperature, the second graph response latency in the Vigilance test, the third detection rate in the same test, and the bottom graph output in the Calculations test.

Temperature rose throughout the shift by a near constant amount each day; the average increase in temperature was of the expected magnitude (cf. Fig. 1). There was also a slight rise in the daily overall mean value of temperature recorded in the early stages of the trial, but this is not thought to be of any significance.

Of the 3 performance measures, Vigilance detection rate showed perhaps the closest relationship to temperature during the shift, the mean score rising considerably between the first and second work-periods, but somewhat less markedly thereafter, with some evidence of post-lunch depression in the third (and also possibly the fourth) period; the trend was, on the whole, reasonably consistent over the 12 days.

Calculations output exhibited a remarkably constant within-shift picture throughout the experiment. There was a clear rise in the first half of the shift, which paralleled the detection-rate increase at this time. There was also post-lunch depression, which was rather more pronounced. However, unlike the detection rate case, there was a consistent recovery from this depression in the fourth work-period. The final Calculations test was, it will be remembered, held later in time in relation to the meal-break than the final Vigilance test, and this may account for the more noticeable recovery in the output score. This recovery itself provides strong support against any argument that cumulative "fatigue" was affecting the performance trends during the shift.

The results for response latency in the Vigilance test were not so clear. Although at the start of the trial, and again at the end, the within-shift changes were very similar to those observed for detection rate and for Calculations output, over the central part of the experimental period there was no clear trend, the changes in latency from work-period to work-period being small and somewhat variable. Nevertheless at no time was

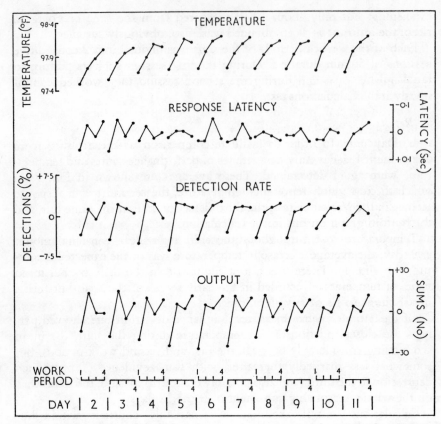

FIG. 9. Eight-hour day shift: 3-day moving averages of mean body temperatures, and of mean performance scores, for each work period of the shift.

there any definite sign of a **downward** trend in on-shift efficiency as measured by this score. Thus, once again, fatigue effects are "not proven".

Separate regression analyses, in which the practice effect was assumed to be linear, were carried out on the results from the first, second, third and fourth quarters of the trial (i.e., Days 1–3, 4–6, 7–9 and 10–12 respectively). These analyses showed

(i) that although the coefficient of regression of **detection rate** on temperature was positive in sign for a majority of the individual subjects' results only in the second and third 3-day periods, the "group mean". coefficient was positive in all quarters of the trial;

(ii) that the coefficient of regression of **Calculations output** on temperature was positive in sign for a majority of the subjects in all quarters but the first (at this early stage of the trial the assumption of linearity in the practice effect for this score may have been incorrect) and

(iii) that despite the rather variable picture presented by the mean scores in Fig. 9, the majority of the coefficients of regression of **response latency** on temperature had the "expected" (i.e. negative) sign in all except the third quarter of the trial.

Thus, in general, it can be concluded that changes in temperature and in performance were related to quite a reasonable extent in this day shift, and that there was no clear evidence that "fatigue" due solely to the length of the duty spell was affecting the results.

(b) Night shift. Means of temperature readings taken over an entire 24-h period on 1 day before the trial began (with normal sleeping hours), on the sixth day, and on the twelfth day are shown in Fig. 10. A distinct change in the rhythm had occurred by the sixth day, and the twelfth day readings suggest that, measured by the degree of "flattening" that was apparent during night hours, this change had progressed somewhat further by the end of the trial. It is difficult to assess the extent to which phase displacement of the rhythm occurred, since there was no clear "peak" in the new curves, but, by inspection, it would appear that on Day 6 there may have been a displacement of about 3 h, and on Day 12

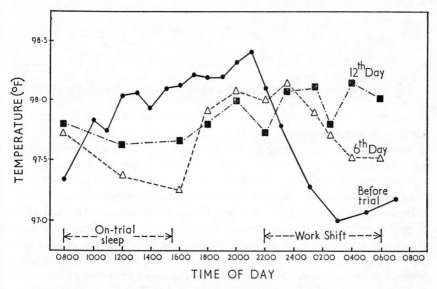

FIG. 10. Eight-hour night shift: mean "24-h" body-temperature curves before, on the sixth, and on the twelfth day of the trial. Reproduced with permission from Colquhoun *et al.* (1968b).

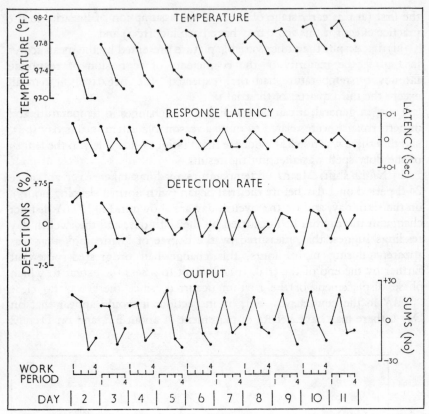

Fig. 11. Eight-hour night shift: 3-day moving averages of mean body temperatures, and of mean performance scores, for each work period of the shift.

one of perhaps 9 h (which would be the "expected" amount if complete adaptation were to occur).

The 3-day moving averages of the within-shift readings (see Fig. 11) indicate that the initial change in the **shape** of the curve during the working period tended to take place on or about the sixth day, but that prior to this the overall daily **mean value** had risen considerably.

Both Vigilance response latency and Calculations output were relatively closely correlated with temperature throughout the trial (Fig. 11). The reversal in temperature trend between work-periods 1 and 2 which took place as part of the overall "flattening" process was reflected in a similar reversal in performance; and the consistent drop in temperature between work-periods 2 and 3 was accompanied by a similarly consistent decline in efficiency. Regression analyses of individual subjects' results for each quarter of the trial showed that, of the 40 coefficients of regression of performance on temperature computed for each score, 33 of those for response latency and 32 of those for output had the "expected" sign. Thus it would appear fairly clear that the within-session trends in these 2 aspects of performance were paralleling that for temperature throughout the whole of the 12-day period. By contrast, Fig. 11 suggests that the on-shift trend in **detection rate** was changing more rapidly than the temperature curve, to the extent that by the eleventh day the initial overall decrement in this score had been transformed into an overall **increment**; in fact, the picture presented by the mean trend for this day resembles to a remarkable degree that shown by the day shift at the beginning of their trial (cf. Fig. 9); even a "post-meal" effect is apparent. This "anticipation" of the temperature change was, it will be remembered, also noted in the detection rate results for Experiment 1. In the present case, all 10 subjects showed evidence of at least partial "performance adaptation" when within-shift trends for the first and last quarters of the trial were compared, and regression analyses revealed that in 5 cases the coefficient of regression of detection rate on temperature was actually negative in sign over the final 3-day period.

This last finding suggests that caution should be exercised in assuming that when a relationship between temperature and performance is found to exist in a "rotating" type of shift system, the same relationship will be found when a change in the temperature rhythm occurs as a result of working a series of night shifts in a "stabilized" type of system. It is interesting to speculate why, in this latter situation as studied here, the scores of Calculations output and of Vigilance response latency appeared to follow the changing temperature trend quite closely, whereas the detection rate index did not. The first two scores, are, of course, both measures of speed, whereas detection rate could perhaps be described as an indicant of "perceptual efficiency". Since there is no overt physical activity involved in the task of merely **detecting** signals, it could be argued that no particular effort is required in doing it, and therefore that the detection rate score would be less subject to fluctuations in, for example, incentive, than would the speed with which a detected signal was actually **reported** (or the rate at which additions were carried out). In this sense it might be said that the detection rate score is a "purer" measure of true mental efficiency than response latency (or Calculation output). If this argument, and the theory that such efficiency is primarily determined by arousal, are both correct, then it follows from the results observed in the present trial that the 24-h "sleepiness" cycle changed at a more rapid rate than the temperature rhythm as the experiment proceeded.

It is possible, then, that the continuing association of temperature and the two "speed" scores throughout the trial was the fortuitous result of the combination of 2 factors that affected these scores in opposing ways. The first of these would be the tendency for speed trends to follow the changing **sleepiness** cycle, and thus to show as much "adaptation" as detection rate; the second would be for speed to fall off during the shift to a progressive extent as the days passed and the level of motivation in the subjects declined. A simpler alternative hypothesis is that speed is in fact related directly to temperature, as Kleitman (1963) holds; however, evidence obtained from the morning shift trial (see below) would seem to favour the former theory to a limited degree.

(c) Morning shift. Means of temperature readings taken over each of 3 selected 24-h periods are plotted in Fig. 12. The curves show that (possibly due to the physical activity involved in rising and in travelling to the laboratory) there was, on the sixth day, an apparent shift of about 1 h in the phase of that section of the rhythm which included the work shift, but that there was no obvious alteration in the overall amplitude of the rhythm. This picture remained essentially unchanged on the twelfth day.

The fact that the changes in the temperature rhythm were very small in this trial meant that any corresponding alterations in performance levels were unlikely to be detectable. Nevertheless it was considered to be of

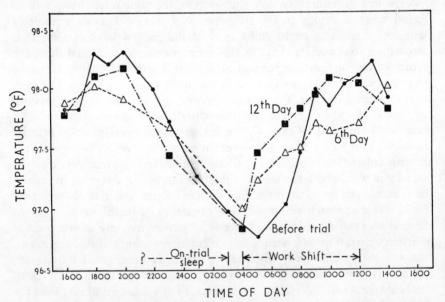

FIG. 12. Eight-hour morning shift: mean "24-h" body-temperature curves before, on the sixth, and on the twelfth day of the trial. Reproduced with permission from Colquhoun *et al.* (1968b).

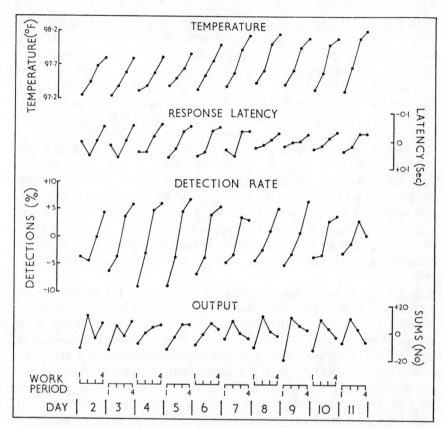

FIG. 13. Eight-hour morning shift: 3-day moving averages of mean body temperatures, and of mean peformance scores, for each work period of the shift.

some interest to examine the trends to determine whether they differed in any general way from those to be expected on the assumption that the relationship between temperature and performance would be essentially similar to that observed in the day shift. Moving averages were therefore computed for each work period in the same manner as before, and the results are shown in Fig. 13.

Although the rise in temperature during this shift, was, as expected, considerably greater than in the day shift, this was reflected only in the detection rate score, variation in which was substantial. However, the overall within-shift improvement in this aspect of performance tended to diminish towards the end of the trial, despite the increasing extent of the within-shift temperature change due to the slight phase displacement of the rhythm over this section of the 24-h period.

D

In the case of response latency the within-shift temperature rise was reflected in an improvement in performance with reasonable fidelity throughout the trial, although, despite the greater range of temperature variation, the daily increase in speed was of no greater magnitude than that found in the day shift. As with detection rate, the extent of this increase in efficiency tended to get smaller towards the end of the trial. This was also true of the Calculations output score, where by the eleventh day the overall within-shift increase in performance level manifested in the earlier stages of the experiment had virtually vanished. Consistent within-shift increase in output throughout the trial was in fact evident only between work periods 1 and 2, and even here the rise was smaller than was the corresponding rise for the day shift. The increasing tendency for output to fall off in work periods 3 and 4, coupled with the less pronounced overall within-shift improvement shown in the Vigilance test scores, suggests that performance in the later stages of this trial was being influenced by factors which were not present in the day shift experiment.

It will be recalled that the subjects assigned to the morning shift experienced considerable difficulty in getting to sleep at the prescribed time of 19.30. It is possible, therefore, that they were suffering from partial sleep deprivation during the trial. If this was indeed the case a reasonable explanation of the decrease in the within-shift rise in performance scores would be that it resulted from the direct effects of this deprivation. On the other hand, the decrease could have been produced by a progressive reduction in motivational level. It is certainly true that 03.00 is a most **unpleasant** time to rise, and so, whether or not sleep was obtained in sufficient quantity beforehand, the repeated experience of being awakened in the middle of the night may very well itself have had an adverse influence on the general morale of the subjects. Some support for this supposition is to be found in the high frequency of complaints about various features of the testing cubicles received from the morning shift groups; such complaints were noticeably absent in the case of the subjects assigned to the day or night shifts.

Thus it would appear that, under "normal" social environmental conditions, commencing work at 04.00 is likely to have a detrimental effect on efficiency in the later part of the shift if the number of consecutive days on which this is done is at all extensive. It is interesting to speculate whether this would still hold true in the special circumstances that would exist in, for example, a spacecraft, or an undersea habitation. In such situations it might be possible to adapt to this particular sleeping–waking routine with relative ease, since there need be no "evening" during which worldly amusements would have to be foregone in order to obtain the necessary amount of sleep. We are assuming here that the primary difficulty in adapting to this routine arises from the conflict between sleep and recreation; it is of course possible that "early" sleep is hard to achieve, not for "social" reasons, but because the high level of body temperature in the late afternoon–evening period tends to interfere directly with the processes involved in its initiation.

## C. *Experiment 3*: *Twelve-hour stabilized systems*

1. *Method.* In this study the length of the duty-spell was increased to 12 h in order to determine the manner in which the previously observed relationships between body temperature and performance were disturbed when the hours of work were extended to the point at which "fatigue" or "stress" arising from the length of the duty-spell would almost certainly itself affect efficiency.

Two 12-h shifts were studied: a day shift (08.00–20.00), and a night shift (20.00–08.00); these 2 shifts represent 1 possible system of continuous manning with 2 operators to each station. The actual times of the shifts in this experimental system were so chosen that the maximum displacement of normal working hours was demanded from the night shift subjects. Thus in this sense the trials could be said to be "testing the limits" as far as possible adaptation to night work was concerned.

Twenty-two ratings volunteered for this study, which, as before, extended over a period of 12 consecutive days. Ten subjects worked the day shift, according to the following routine:

| Time | Activity |
|------|----------|
| 08.00–09.40 | Work period 1 |
| 09.40–09.50 | Tea break |
| 09.50–11.30 | Work period 2 |
| 11.30–12.30 | Meal break |
| 12.30–14.10 | Work period 3 |
| 14.10–14.20 | Tea break |
| 14.20–16.00 | Work period 4 |
| 16.00–16.30 | Snack break |
| 16.30–18.10 | Work period 5 |
| 18.10–18.20 | Tea break |
| 18.20–20.00 | Work period 6 |

Sleep time in this trial was from 23.00 to 06.30, and, as in Experiment 2, subjects were excused any additional duties outside their working hours. Body temperature was recorded at 08.00, 09.40, 11.30, 12.30, 14.10, 16.00, 18.10 and 20.00.

The within-shift routine for the night shift, worked by 12 subjects, was identical with that for the day shift, apart from the 12-h time displacement. Sleep time for 6 of the subjects on this shift was from 10.30 to 17.30; for the other 6 it was from 11.30 to 18.30. Thus not only the duty spell, but the entire 24-h activity cycle for these night shift subjects was displaced by approximately 12 h.

Six subjects in each shift performed the Vigilance test during each of the 6 work periods; the test was modified to enable the stimuli to be displayed visually as well as aurally for this experiment. The test was presented for one third of the work period in this new visual form, for a second third in the original auditory form, and for a third in a combined auditory–visual form. The order in which the 3 display forms were presented was permuted in successive work periods.

The remaining subjects in each shift performed the Vigilance test only in alternate work periods; in the remainder they were given the Calculations test. The sequence in which the 2 tasks were carried out was reversed on successive days. During "vigilance" work periods the display changed from the auditory to the visual form (or vice versa) at the half-way mark, the order of presentation of the 2 forms being reversed in successive sessions. The combined audio–visual form of presentation was not used with these subjects.

## 2. Results.

(a) Day shift.

(i) "Vigilance only" group. Since no obvious difference was apparent in the within-shift performance trends obtained with auditory, visual, and audio–visual forms of the task, the data were collated over all three display conditions. Three-day moving averages of the resulting scores, and the corresponding temperature readings, are shown in Fig. 14.

As expected, mean temperature rose throughout the shift on each day, following its natural rhythm. Although there was a correspondingly consistent **overall** rise in detection rate during the shift, the scores in successive work periods were not a simple monotonic function of these periods; the characteristic trend was a sharp rise between the first and second periods, followed by a fall in the post-lunch phase, and a subsequent second rise to a peak in the final work period. The post-lunch effect appeared to extend, particularly in the later stages of the trial, as far as the fourth work period, i.e., for as long as 3 h after the meal break.

As might be expected from this disparity between the form of the mean within-shift trends in temperature and performance, the coefficients of regression of detection rate on temperature computed for individual subjects in successive quarters of the trial varied somewhat both in sign and in reliability. Nevertheless, in each 3-day period except the second the **mean** coefficient was positive.

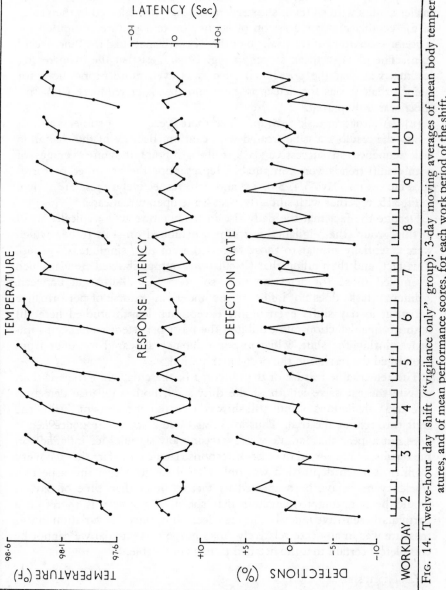

FIG. 14. Twelve-hour day shift ("vigilance only" group): 3-day moving averages of mean body temperatures, and of mean performance scores, for each work period of the shift.

The degree of within-shift variation in response latency was very small, and the overall relationship between temperature and performance during the day was, if anything, the reverse of that predicted on the basis of the earlier results with shifts of shorter length (this was reflected in the values of the coefficients of regression of latency on temperature computed for successive quarters of the trial; 16 of the 24 coefficients had the "incorrect" sign). The plots of mean scores in Fig. 14 suggest that the main reason for this was that the post-lunch effect was even more pronounced for latency than it was for detection rate, and that later recovery from this effect was only partial.

(ii) "Alternating task" group. Since there were only 4 subjects in this group the results must be treated with caution. Because of the change in task sequence on successive days, only 4 3-point moving averages of within-shift trends were computable, representing the smoothed performance curves on Days 3/4, 5/6, 7/8 and 9/10 respectively; these are shown in Fig. 15, together with similarly assessed temperature means.

Despite the fact that the daily rise in temperature was atypically small in this group, the Vigilance test scores exhibited on-shift trends which were essentially similar to those recorded from the "single-task" group. This fact, and the finding that Calculations output showed trends which resembled those for detection rate, suggests that alternation between 2 different tasks does not influence the underlying course of performance variation in day shifts even of the exceptional length studied here. In both groups detection rate exhibited the most persistent tendency to improve during the shift, while response latency appeared to suffer from prolonged depression in the post-lunch period.

The remarkable feature of the outcome of these day shift trials is that, despite the excessive length of the duty-spell, both Vigilance detection rate and Calculations output still showed overall improvement during the shift throughout the trial. Thus as far as these 2 scores are concerned it does not appear that fatigue was exercising any significance influence on the results. However, in the case of response latency, the fact that recovery from post-lunch depression was only partial suggests that this aspect of efficiency may have been affected by factors other than time of day. It would be tempting to conclude that speed of response is therefore a particularly sensitive index of fatigue effects, but, since the variation in this measure was in any case relatively small in the present trial we cannot be completely certain that the detailed trends are reliable.

(b) Night shift.

(i) "Vigilance only" group. Means of "24-h" temperature readings taken before, and twice during the trial period are shown in Fig. 16 for this

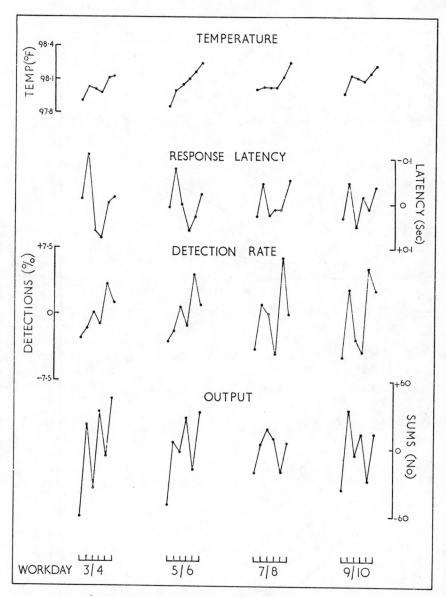

FIG. 15. Twelve-hour day shift ("alternating task" group): 3-day moving averages of mean body temperatures, and of mean performance scores, for each work period of the shift.

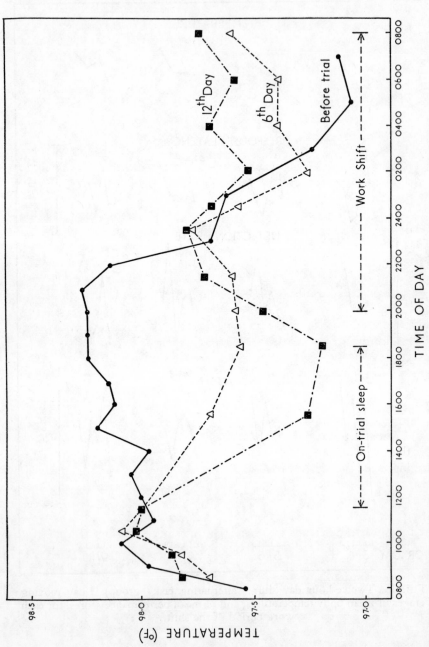

FIG. 16. Twelve-day night shift ("vigilance only" group): mean "24-hour" body-temperature curves before, on the sixth, and on the twelfth day of the trial.

"single-task" group. The curves suggest that a change in the rhythm had occurred by the sixth day, where there was some evidence of a 12-h periodicity, with peaks at roughly 10.30 and 23.30. The latter peak was about 2 h later than the "normal" peak; the former was displaced from it by approximately 10 h. The most marked feature of the new curve, however, was that its amplitude was considerably less than the original one, i.e. "flattening" had occurred. Little change was noted in this picture on the twelfth day, except for a suggestion that the change had proceeded a little further in respect of the degree of "flattening" that was present during the work shift, and also in the depth of the sleep-period trough.

Notwithstanding the fact that the temperature "adaptation" was only partial, and also apparently less marked than in the 8-h night shift trial in Experiment 2, the change that occurred was sufficiently large to effect a substantial reduction in the extent of the overall decline in temperature during the working period. Three-day moving averages of the actual readings (see Fig. 17) suggest that, as in both earlier experiments with night shifts, the initial change in within-shift temperature trend tended to take place on or about the sixth day.

Response latency was fairly well related to temperature throughout the trial in this group of subjects. The reduction in the extent of the fall in temperature during the shift consequent on the flattening of the rhythm was accompanied by a corresponding reduction in the fall in speed. On the other hand, no clear change in the within-shift detection rate trend appeared to occur until the last 3 days of the trial—a finding which contrasts with those from the night shift trials in Experiments 1 and 2, in both of which the changes in this score seemed to "anticipate" those in the temperature readings.

(ii) "Alternating task" group. The "24-h" temperature curves recorded from this group of subjects (Fig. 18) suggest that the change in the rhythm resulting from the abnormal routine was smaller than in the "vigilance only" group. A 12-h periodicity was again evident, with peaks at about the same times as previously observed, but the "flattening" of the rhythm during the work shift was considerably less marked. Although it is true that there was a difference of 1 h in the times between which sleep was taken in the 2 groups, it is not thought that this could have accounted for the difference in the temperature curves obtained. In default of any other reasonable explanation, this must be ascribed to the inclusion, by chance, of a majority of "good adapters" in the first group, and of "poor adapters" in the second.

The fact that flattening of the on-shift temperature curve was very slight in the "alternating task" group would lead one to expect that

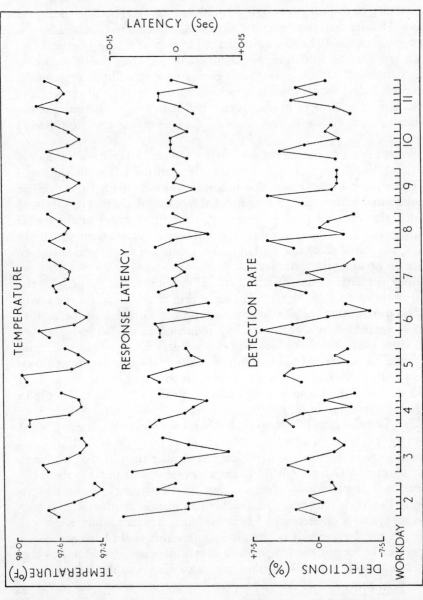

FIG. 17. Twelve-day night shift ("vigilance only" group): 3-day moving averages of mean body temperatures, and of mean performance scores, for each work period of the shift.

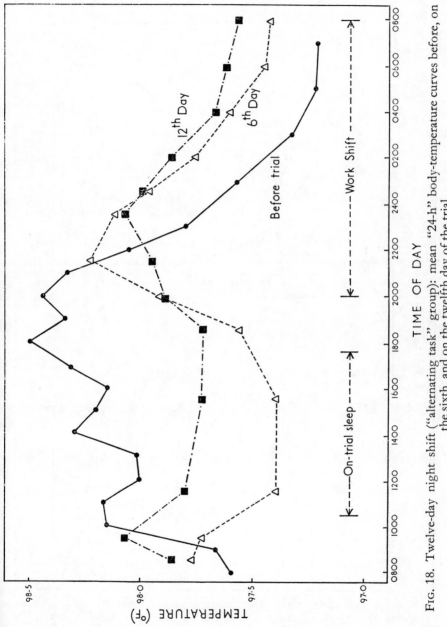

FIG. 18. Twelve-day night shift ("alternating task" group): mean "24-h" body-temperature curves before, on the sixth, and on the twelfth day of the trial.

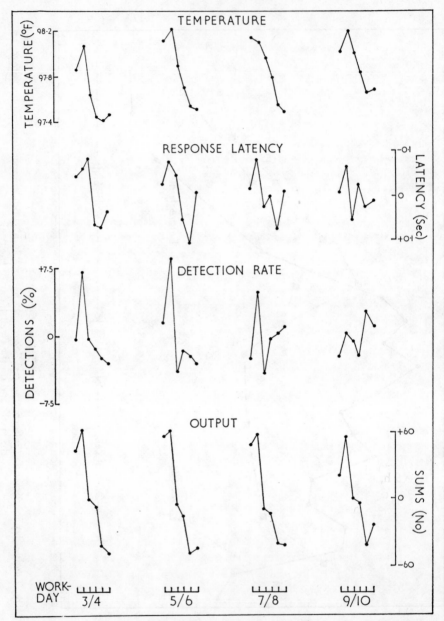

Fig. 19. Twelve-day night shift ("alternating task" group): 3-day moving averages of mean body temperatures, and of mean performance scores, for each work period of the shift.

changes in performance trends would be correspondingly small, if detectable at all. The moving averages plotted in Fig. 19 show that this was indeed the case for the scores of response latency and Calculations output, both of which followed the downward trend of temperature which persisted throughout the trial. However, detection rate **did** show evidence of "adaptation" at the later stages of the 12-day period, much as it did in the "vigilance only" group. Thus it would appear that, as in the case of day work, task alternation makes no difference to the pattern of performance trends obtained on successive 12-h shifts.

The finding that the within-shift trend in the Vigilance detection rate score altered in the later stages of this trial in **both** groups of subjects, whereas the response latency curve changed only in that group where the temperature rhythm showed definite evidence of flattening, supports our earlier contention that detection rate may be a more sensitive index of "true" efficiency than speed of reaction in this test (or rate of work in the Calculations test). Once again, assuming the correctness of the "arousal" theory of performance, the implication is that, when abnormal routines are followed, alterations in the 24-h sleepiness cycle take place independently of changes in the temperature rhythm. It was argued from the results of the night shift trial in Experiment 2 that the rate at which such alterations occurred was faster than the temperature curves would indicate. The results obtained from the "alternating task" group in the present night shift trial could be said to be generally consistent with this view, in that a clear alteration in the detection rate trend took place despite the fact that there was very little change in the temperature rhythm; however, the actual stage in the trial at which this alteration appeared was considerably later than in the previous case. A possible explanation for this delay in apparent adaptation of the sleepiness cycle may be that it takes longer to adjust to a 12-h displacement of the sleep period than it does to one of only 9 h, as the schedule in Experiment 2 required.

The data from the "vigilance-only" group are in one sense consistent with this concept of delayed adaptation due to a maximum displacement of sleep hours, since the change in the detection rate trend in this case also took place later in the trial than it did in Experiment 2. However, in this group the temperature curve showed clear evidence of flattening several days **before** the detection rate change; thus the results here are anomalous as far as the relative rates of change of temperature and sleepiness rhythms are concerned. Although several explanations might be advanced for this difference between the groups, these could at best be only speculative, since the numbers of subjects providing the scores on which the plots of average trends in Figs 17 and 19 are based were relatively small, and there are undoubtedly substantial individual differences in all these matters. Thus the somewhat overworked conclusion "further research is needed" would seem in fact to be the only appropriate one at this stage.

## V. Concluding Remarks

It is clear from the foregoing account that the efficiency with which many kinds of mental task are carried out does in fact vary considerably, and in a systematic manner, according to the time of day or night at which the measurements are taken. However, it is also evident that a great deal of work remains to be done before a full understanding of this circadian periodicity is achieved. In this section some of the areas to which further research might profitably be devoted will be outlined.

The range of questions to which a future research programme might attempt to find answers is very large. Even in the restricted field of "waking day" studies, for example, information is badly needed on the influence of a whole range of factors on the "normal" performance rhythm. Some of these, such as the duration and quality of previous sleep, the physical health of the subject, and the presence of stimulant or depressant drugs in the bloodstream, can be said to be of a basically "physiological" nature. Others, including the degree of intelligence, the level of motivation, and the personality structure of the individual are more obviously "psychological", as is the extent of social interaction permitted during the observation period. A third group, which cannot be readily classified in this way, covers such diverse variables as the nature of the environmental conditions under which the tasks are performed (e.g. quiet or noisy, cool or hot), the timing and content of meals, the frequency of rest breaks, the length of experience with **abnormal** living routines, and the season of the year. Whether the rhythm alters with age, and whether it is the same in women as in men are other basic questions to which an answer should be sought. Again, the form (or even the existence) of rhythms in cognitive or "higher-level" functions such as memory and decision-making still requires to be established; the premium on constant efficiency at this kind of mental activity will inevitably rise with the new skills required by the operators of automatic plant and machinery.

A separate set of problems exists in the wider area of "round-the-clock" studies. Further work is urgently needed here, as the proportion of the population which must come to accept shift working as a normal way of life will in all probability rise considerably in the next few years. Perhaps the greatest difficulty with shift working is the disturbance of the normal sleep–waking pattern on "night" shifts. Although it seems well established that, on average, the rate of adaptation to night work is very slow, it would appear that there are some people whose rhythms change more rapidly in response to unusual sleep–waking routines than others; intensive study of the physiological and psychological constitutions of these individuals might lead not only to a deeper understanding of the role of *Zeitgebers* in controlling periodicity, but also to the development of methods for selecting suitable workers for night time operations. Should the number of such persons be found to be too small to meet the demand, it might prove necessary to attempt to formulate a technique of "training" specifically for night work. This assumes that the main requirement will be for people who are able (and willing) to work at least a "5-night" week, if not a longer one. However, the growing popularity of more rapidly rotating shift-systems such as the "Continental" (2–2–3) arrangement suggests that social pressures may result in systems of this kind becoming more widely adopted in the future. If this is the case, the emphasis will then tend towards intensive examination of performance during the first stages of periods of inverted sleep–waking cycling, and the importance of the timing, duration, and above all the **quality** of day-time sleep in these stages may become the focus of attention.

It is almost inevitable that the effects of circadian periodicity on the performance of people whose sleep–waking routine is inverted by a spell on night-shift will, on the first night, be completely confounded with the effects of ongoing sleep deprivation. It is also highly probable that, on the second and subsequent nights, the on-shift readings will be influenced by the effects of **partial** and/or **qualitative** sleep deprivation as a result of the reduced, disturbed, or broken sleep that typically occurs when sleep is taken during daylight hours (Tune, 1968). Furthermore, it is entirely possible that the rate at which various rhythms alter is itself affected by different amounts, kinds and patterns of daylight sleep. If research proves this to be so, further experiments could be mounted to determine the feasibility of accelerating this rate of alteration by the use of appropriate sleep-inducing drugs.

The main concern in shift-work studies has been with the difficulties experienced in changing from a "normal" to an "abnormal" routine; little attention has been paid to the other side of the coin, i.e., the rate of **re-adaptation** after the night-work period ends. This is presumably because a rest-interval of about 2 days is usually programmed between the end of the night shift period and the beginning of the next spell of duty, and it is generally assumed that this 2-day interval is sufficiently long for "recovery" to take place. This assumption may or may not be a valid one; however, in rapidly rotating systems there is the danger that the intershift rest interval may be reduced, or in some cases omitted altogether. A pertinent topic for research, then, is to discover how long it does in fact take for disturbed rhythms to re-adapt either to a "normal" routine or to a different abnormal one. A related question is the magnitude of the "step" or phase-shift that can be "absorbed" by the system without disruption; it is possible, for instance, that were the sleep-waking routine to be advanced or delayed in short daily steps of, say, 60 min, adaptation of the rhythms might occur with very little, if any time–lag. If this was indeed the case, a consistently high level of efficiency at all hours of the day and night could at any rate theoretically be achieved in a given factory by continuously altering the sleep–waking routine of 3 teams of shift workers in a similar way; each team would then live permanently on a 23-h or 25-h "day" (cf. Eranko, 1957) and carry out their duty-spell at the optimum time within this "day".

Experiments like these on the speed with which rhythms can be entrained will no doubt be carried out initially in the laboratory, but it is essential that the results obtained from them, and indeed from all other experiments in this area, should be carefully validated in actual operational situations. The problems met with in field studies of circadian rhythms are formidable; not least among them is the difficulty of gaining access to the subjects, who are, of course, in this case normally fully occupied with their own work. There may be a very few situations where this work is iteself of such a nature as to provide direct measures of "true" efficiency, but in the majority of cases the latter will have to be assessed by interpolated tests. Thus one of the major initial concerns of a laboratory research programme should be the creation of tests which, though short, are nevertheless particularly sensitive to circadian fluctuations, simple to administer, and also of high reliability. Such tests would not only facilitate data collection in the field, but would also provide the means by which a systematic search could be made for rhythms of shorter period than the circadian, e.g., the hypothetical "90-min" cycle suggested earlier in this chapter as one explanation for the post-lunch phenomenon.

Along with these improvements in the design of the basic tools of research should go a parallel effort in developing tests which enable assessments to be made of variations in the underlying mental processes responsible for the observed performance scores which are the "raw data" of experimental results. An example of this sort of development can be found in the history of the vigilance test used by the present author in the experiments described earlier. The original form of this test measured only simple detection rate; however, progressive refinements in scoring techniques led to the present version, with which it is possible to distinguish genuine changes in discriminatory efficiency from those due to alteration in response biases. Tests of higher level functions such as memory and decision-making should provide even greater scope for the exercise of such an "analytical" approach.

A final, but not least important "methodological" issue requiring attention is the question of how the data gathered both in laboratory and field experiments can best be analysed to determine the existence of periodicities. Mention has already been made of this problem earlier, and a comparison of the description of available mathematical techniques described in Chapter 1 with the analyses actually carried out on the present author's results provides a good example of the gap between theory and application. The bridging of this gap should be approached from both ends, since whereas it is clear that at least some of the people working on circadian rhythms in performance badly need educating in the mathematical skills of time series analyses, it is also true that the typical models proposed by the mathematicians themselves are ill-equipped to deal with the kind of data that studies of rhythms in human subjects tend to produce. For unavoidable reasons such data is often incomplete, and it is also not normally practicable to ensure that the frequency and regularity of a series of experimental observations matches up to the requirements laid down by theoreticians. Add to this the possibility of non-stationarity in successive cycles, and the near certainty of substantial "practice" effects, and it becomes clear that an analytical technique which will be successful in reliably determining the existence of periodicities in the sort of data which are usually obtained in circadian performance studies must possess a high degree of sophistication. Nevertheless, the development of such a technique is probably as important to progress in the understanding of the nature of periodicity in mental efficiency as is the selection of the most appropriate experimental programme for the purpose.

Since mathematicians have now been brought into the picture, this is a convenient point at which to stress that, because of the increasing sophistication of research techniques, a multi-disciplinary approach to the future study of circadian rhythms is not only inevitable, but indeed almost essential if worthwhile results are to be achieved. In the past this has not always been recognized. For example, until recently there has been an almost total lack of studies of rhythms in which psychologists and physiologists have co-operated. It is most probable that this unfortunate division stems from the historical development of psychology in the universities, where it has almost always been regarded as an "arts" subject. Now that psychology has acquired scientific respectability, progress demands that the two disciplines collaborate in future research in this area, as indeed they are already doing in others. Given the necessary financial backing (and there should be many interested parties in industry and government) the way would then be open for new-style investigations on a scale which hitherto has not seriously been contemplated. Some at least of these investigations would be conducted in isolation suites specially constructed to exclude all naturally occurring *Zeitgebers*. In such suites studies could be made of the adaptability of many different rhythms, both physiological and psychological, to changes in the sleep–waking routine that correspond to those to which shift-workers, and people subject to multiple time-zone transitions, are actually exposed. The results of these studies would, of course, have obvious and immediate applications to existing real-life problems in these areas, but, as the research programme developed, facilities such as described above could also be used to advantage to examine the effects of alterations in the sleep–waking cycle which are normally outside the experience of man in his natural environment. These investigations might lead, not only to greater insight into the underlying mechanism of periodicity in organismic functioning, but also to optimal solutions for the future problems of mass space-travel and extensive underwater habitation where the normal *Zeitgebers* are not present.

The long-term aim of the research programme that has been outlined would be to identify the fundamental determinants of circadian rhythms in human performance. In this chapter it has been suggested that these rhythms are related to fluctuations in the "state of arousal" of the organism. The physiological concomitants of the arousal state may well prove to be extremely complex, and no one measured variable may be found adequate for reflecting variations in it with sufficient consistency. Thus, as we have seen, the earlier view that body temperature might serve this purpose has been called into question by some of the more recent findings (e.g. on the post-lunch phenomenon, and the differential rates at which performance and temperature rhythms alter in response to a changed sleep–waking routine). Nevertheless it remains true that a **direct** test of the dependence of mental efficiency on the body-temperature rhythm has never been made. In an appropriately equipped laboratory it would be possible to make such a test by deliberately interfering with the normal temperature cycle (perhaps through the medium of anti-pyretic drugs) in order to alter its characteristics. This kind of manipulative technique could be extended to hormonal and other factors exhibiting circadian periodicity, this periodicity being either eliminated or altered systematically during the course of a regular series of performance–testing observations in order to determine the existence of any dependency between physiological and psychological parameters in each case.

Should it be found that such dependencies were not demonstrable in the case of any "vegetative" process, and that there were self-sustaining periodicities in performance, two conclusions might follow. It might be that the "state of arousal" of the organism (or possibly some more specific parameter) is independently oscillatory in a way that is determined by some innate organization of the brain. Alternatively, it is possible that time dependencies are learned, and the implication that circadian variations in performance are based on the achievement of a stereotype of time-conditioned reflexes would then have to be explored.

## References

Aarons, L. (1968). *Psychophysiology* **5**, 77–90.
Adkins, S. (1964). *Percept. Mot. Skills* **18**, 409–412.
Alluisi, E. A. and Chiles, W. D. (1967). *Acta Psychol.* **27**, 436–442.
Aschoff, J. (1960). *Cold Spring Harbor Symp. Quant. Biol.* **25**, 11–28.
Aschoff, J. and Wever, R. (1962). *Naturwissenschaften* **49**, 337–342.
Baddeley, A. D. (1966). *Amer. J. Psychol.* **79**, 475–479.
Benedict, F. G. and Snell, J. F. (1902). *Pfluegers Arch. Gesamte Physiol. Menschen* **90**, 33–72.
Bjerner, B., Holm, A. and Swensson, A. (1955). *Brit. J. Ind. Med.* **12**, 103–110.
Blake, M. J. F. (1967). *Psychon. Sci.* **9**, 349–350.
Bogoslovsky, A. I. (1937). *Bull. Biol. Med. Exp* (URSS) **3**, 127–129.
Brown, I. D. (1967). *Ergonomics* **10**, 665–673.
Browne, R. C. (1949). *Occup. Psychol.* **23**, 1–6.
Burton, A. C. (1956). *Can. Med. Ass. J.* **75**, 715–720.
Chiles, W. D., Alluisi, E. A. and Adams, O. S. (1968). *Hum. Fact.* **10**, 143–196.
Colquhoun, W. P. (1962). *Bull. du. C.E.R.P.* (Paris) **11**, 27–44.
Colquhoun, W. P., Blake, M. J. F. and Edwards, R. S. (1968a). *Ergonomics* **11**, 437–453.
Colquhoun, W. P., Blake, M. J. F. and Edwards, R. S. (1968b). *Ergonomics* **11**, 527–546.
Colquhoun, W. P., Blake, M. J. F. and Edwards, R. S. (1969). *Ergonomics* **12**, 865–882.
Corcoran, D. W. J. (1962). "Individual Differences in Performance after Loss of Sleep". Unpublished Ph.D. Thesis, University of Cambridge, England.
Crozier, W. J. (1926). *J. Gen. Physiol.* **9**, 531–546.
Drucker, E. H., Cannon, L. D. and Ware, J. R. (1969). *Human Resources Research Office* (*U.S.A.*) Report No. 69–8.
Duffy, E. (1962). "Activation and Behaviour". Wiley, New York.
Eiff, A. W. von, Bockh, H., Gopfert, F., Pfleiderer, F. and Steffen, T. (1953). *Z. Gesamte Exp. Med.* **120**, 295–307.
Eranko, O. (1957). *Int. Congr. Occup. Health* **3**, 134.
Fiorica, V., Higgins, E. A., Iampietro, P. F., Lategola, M. T. and Davis, A. W. (1968). *J. Appl. Physiol.* **24**, 167–176.
Fort, A. (1968). *J. Physiol.* (*London*) **197**, 82–83.
Fox, R. H., Bradbury, P. A., Hampton, I. F. G. and Legg, C. F. (1967). *J. Exp. Psychol.* **75**, 88–96.
Francois, M. (1927). *L'Annee Psychol.* **28**, 186–204.
Frazier, T. W., Rummel, J. A. and Lipscomb, H. S. (1968). *Aerosp. Med.* **39**, 383–395.
Freeman, G. L. (1948). "The Energetics of Human Behaviour". Cornell University Press, Ithaca, N. Y.
Freeman, G. L. and Hovland, C. I. (1934). *Psychol. Bull.* **31**, 777–799.
Gates, A. I. (1916). *Univ. Calif. Publ. Psychol.* **1**, 323–344.

Gavrilescu, N., Pafnote, M., Vaida, I., Mihaila, I., Luchian, O. and Popescu, P. (1967). *Fiziol. Norm. Patol.* **13**, 421–427.

Gibson, R. B. (1905). *Amer. J. Med. Sci.* **129**, 1048–1059.

Guilford, J. P. (1956). "Fundamental Statistics in Psychology and Education". McGraw-Hill, New York.

Halberg, F. (1965). *In* "Circadian Clocks" (J. Aschoff, ed.), pp. 13–22. North-Holland Publishing Co., Amsterdam.

Hauty, G. T. (1962). *Ann. N.Y. Acad. Sci.* **98**, 1116–1125.

Hauty, G. T. and Adams, T. (1965). *In* "Circadian Clocks" (J. Aschoff, ed.), pp. 413–424. North-Holland Publishing Co., Amsterdam.

Hoagland, H. (1933). *J. Gen. Psychol.* **9**, 267–287.

Jores, A. and Frees, J. (1937). *Deut. Med. Wochenschr.* **63**, 962–963.

Kleber, R. S., Lhamon, W. T. and Goldstone, S. (1963). *J. Comp. Physiol. Psychol.* **56**, 362–365.

Klein, K. E., Bruner, H., Holtmann, H., Rehme, H., Stolze, J., Steinhoff, W. D. and Wegmann, H. M. (1970). *Aerosp. Med.* **41**, 125–132.

Klein, K. E., Wegmann, H. M. and Bruner, H. (1968). *Aerosp. Med.* **39**, 512–518.

Kleitman, N. (1963). "Sleep and Wakefulness". University of Chicago Press, Chicago.

Kleitman, N. and Jackson, D. P. (1950). *J. Appl. Physiol.* **3**, 309–328.

Kleitman, N., Titelbaum, S. and Feiveson, P. (1938). *Amer. J. Physiol.* **121**, 495–501.

Lehmann, G. (1953). *Acta. Med. Scand.* Suppl. **278**, 108–109.

Lobban, M. C. (1965). *In* "The Physiology of Human Survival" (O. G. Edholm and A. L. Bacharach, eds). pp. 351–386. Academic Press, New York and London.

Lockhart, J. M. (1967). *J. Exp. Psychol.* **73**, 286–291.

van Loon, J. H. (1963). *Ergonomics* **6**, 267–273.

Loveland, N. T. and Williams, H. L. (1963). *Percept. Mot. Skills* **16**, 923–929.

de la Mare, G. and Walker, J. (1968). *Occup. Psychol.* **42**, 1–21.

Menzel, W. (1962). "Menschliche Tag-Nacht-Rhythmik und Schichtarbert". Benno Schwabe, Basel.

Mills, J. N. (1966). *Physiol. Rev.* **46**, 128–171.

Murray, E. J., Williams, H. L., and Lubin, A. (1958). *J. Exp. Psychol.* **56**, 271–273.

Murrell, K. F. H. (1965). "Ergonomics". Chapman and Hall, London.

Oginski, A. (1966). *Proc. XV Int. Congr. Occup. Med.* **4**, 95–98.

Pfaff, D. (1968). *J. Exp. Psychol.* **76**, 419–422.

Rubenstein, L. (1961). *Science* **134**, 1519–1520.

Rutenfranz, J. and Helbruegge, T. (1957). *Z. Kinderheilk.* **80**, 65–81.

Sharp, G. W. G. (1961). *Nature (London)* **190**, 146–148.

Swets, J. A. (ed.) (1964). "Signal detection and recognition by human observers". Wiley, New York.

Teleky, L. (1943). *Ind. Med.* **12**, 758–779.

Thompson, C. (1967). "An Investigation of the Daily Activity in the Normal Electroencephalogram, and its Relation to Physiological and Psychological Circadian Rhythms". Unpublished M.Sc. thesis, University of Aston, England.

Thor, D. H. (1962a). *Percept. Mot. Skills*, **15**, 451–454.

Thor, D. H. (1962b). *Psychol. Record*, **12**, 417–422.

Toulouse, E. and Pieron, H. (1907). *J. Physiol.* (Paris) **9**, 425–440.
Trumbull, R. (1966). *Hum. Fact.* **5**, 385–398.
Tune, G. S. (1968). *Sci. Journal*, 67–71.
Ulich, E. (1964). "Schicht und Nachtarbeit im Betrieb". Westdeutscher Verlag, Koln und Oplagen.
Verkhutina, A. I. and Efimov, V. V. (1947). *Byull Eksp. Biol. Med.* **23**, 37–40.
Viteles, M. S. (1932). "Industrial Psychology". W. W. Norton, New York.
Wilkinson, R. T. and Edwards, R. S. (1968). *Psychon. Sci.* **13**, 205–206.
Wittersheim, G., Grivel, F. and Metz, B. (1958). *C.R. Soc. Biol.* **152**, 1194–1198.
Wyatt, S. and Marriott, R. (1953). *Brit. J. Ind. Med.* **10**, 164–172.

CHAPTER 3

# Temperament and Time of Day

M. J. F. BLAKE*

*MRC Applied Psychology Unit, Cambridge, England*

## I. Introduction

*The evidence presented in the previous chapter demonstrated that, in many conditions, performance at certain kinds of task exhibits a quite marked circadian periodicity, which appears, in general, to be in phase with the concurrent rhythm of body-temperature. The present chapter describes in some detail the results of one particular series of experiments in which the relationship between variations in temperature and performance during the waking day was studied over a wide range of "mental" tasks; and attempts to relate the individual differences observed to an objective measure of a particular personality characteristic, namely, introversion–extraversion. This series of experiments was referred to in the "waking day" section of the previous chapter.*

*As already noted, a common shortcoming of the majority of studies on time of day effects on performance has been the relatively small number of subjects observed in any single investigation. In "long-term" experiments such as the shift-work and watch-keeping projects described in the last chapter this defect is understandable, since the magnitude of the undertaking almost invariably precludes the use of a larger number of individuals. The same restriction does not necessarily apply to "short-term" experiments, however, and one of the aims of the investigation to be described here was to ensure that the number of individuals tested in each case was*

---

* This chapter has been compiled by the Editor from a partial draft prepared by the author before his recent accidental death. The Editor takes full responsibility for any errors or omissions resulting from this undertaking, in which he has been greatly assisted by many colleagues, particularly P. M. E. Altham, D. W. J. Corcoran, and P. R. Freeman. A distinction between that part of the text which is based on the author's original contribution and that which consists purely of editorial comment has been drawn by setting the latter in italic type.

*sufficiently large to enable relatively firm conclusions to be drawn about any effects observed. It is obvious also that the second aim of the present study, the analysis of individual differences, would in any case hardly be practicable with very small groups.*

Since previous studies on circadian periodicity in performance were reviewed in the last chapter, it is only necessary here to mention those particular investigations in which individual differences have been discussed in any detail. Unfortunately the number of such cases is very small indeed, despite the fact that, as Kleitman (1963) reminds us "the existence of distinct 'morning' and 'evening' types is . . . an everyday observation". Kleitman himself holds that "there are 2 distinct types of body-temperature and efficiency curves, with the peak reached early in the waking period in one and later in the other. In addition, there are intermediate gradations between the 2 extremes" (Kleitman, 1963, p. 161). By way of example, he quotes results from 2 of his own subjects, both of whom showed close correlations between hand-steadiness and temperature as the latter varied during the day, but with clearly different maxima. In further studies, using both simple and choice reaction-time tests, he found that, in 6 subjects "it would appear that $RT$ was always connected with the body temperature" (Kleitman, 1963, p. 154). This implies that the differences in temperature rhythm in the different subjects were reflected also in the differences in the $RT$ rhythm.

Although these results are encouraging, it is nevertheless true, as Kleitman points out, that practically no other investigator other than Kleitman himself has recorded body-temperature when seeking for individual differences in diurnal variation in performance, or, for that matter, has recorded performance when assessing the range of diurnal variation in temperature. This is no doubt due to the (hitherto) disparate interests of psychologists on the one hand, and of physiologists on the other.

When we consider possible associations of differences in temperature or performance rhythms with other characteristics of the subjects, such as intelligence or personality, there are only 2 relevant studies known to the present writer. Colquhoun (1960) reported significant individual differences in performance at a prolonged paced inspection task in morning and afternoon sessions; efficiency at detecting small changes in visual stimuli was **positively** correlated with a measure of introversion in the morning, but **negatively** correlated with the same measure in the afternoon. Unfortunately, different groups of subjects were used at the 2 test-times and body temperature was not recorded, so it was not possible to examine the relationship between physiological and psychological rhythms. In a follow-up of this study (again using different groups, and again without recording temperature) Colquhoun and Corcoran (1964) demonstrated that, at least for "morning" tests, the relationship between performance and introversion–extraversion was not specific to the original inspection task. In this experiment the task was "letter cancellation", and unpaced; rate of work in a 15-min session was found to be significantly higher in "introverts" than in "extraverts" at 08.30. At 13.30 this finding was reversed, but the difference in speed was small and not statistically significant (this may possibly have been due to confounding by the "post-lunch" effect already referred to in the previous chapter). Error rate in this task was apparently unrelated to introversion–extraversion at either of the 2 test times.

There are 3 points about the study by Colquhoun and Corcoran which should be noted. The first is that, when subjects were tested together as a group the relationship between introversion and performance disappeared. The second is that with this unpaced task it was speed, rather than accuracy, which was the sensitive variable; and the third is that (unlike the inspection task) the test was a relatively short one. Thus it would appear that whereas in experiments in this area the degree of isolation of the subject must be carefully controlled, it is possible to use tests which are either paced or unpaced, and of short or long duration.

In summary, the position on individual differences in circadian rhythms before the present experiments were undertaken appeared to be as follows: first, there was a considerable amount of evidence that, whereas almost all people exhibit a 24-h fluctuation in body temperature, the maximum level in this cycle occurs at varying times, normally ranging from mid-day to late evening; second, there was some indication that these differences in cycle phase are reflected in correlated performance efficiency rhythms; and third, there was a strong suggestion that individual differences in performance at particular times of day are related to the introversion–extraversion dimension of personality.

A new piece of relevant evidence should now be mentioned. This was an analysis by the present author (Blake, 1967) of the normal circadian rhythm of body temperature in a group of 74 Naval ratings, most of whom took part in the "watch-keeping" experiments described in the previous chapter. For this analysis, sub-lingual temperatures were recorded 20 times during a single 24-h period, on 2 occasions separated by approximately 1 week. Readings were taken hourly between 07.00 and 23.00, and 2-hourly between 23.00 and 07.00 (the subjects' sleep period). Body temperature at each of the 20 times of day was taken as the mean of the 2 readings obtained. These mean body temperatures were then correlated with the subjects' scores on the "unsociability" scale of the Heron Personality Inventory (Heron, 1956), which was the same measure of introversion–extraversion used by Colquhoun in his studies described above. It was found that the correlation changed from significantly positive (introverts with relatively **higher** temperatures) to significantly negative (introverts with relatively **lower** temperatures) over the period from 08.00 to 21.00. During the late evening and the sleeping hours this trend was reversed (see Table I).

TABLE I

*Correlation coefficients* (r) *of body temperature and introversion rating at twenty times o, day* (n = 74). *Reproduced with permission from Blake (1967)*

| Time | | | | | | |
|------|---------|---------|---------|----------|----------|---------|
| Time | 05.00 | — | 07.00 | 08.00 | 09.00 | 10.00 |
| r | −0·053 | — | +0·133 | +0·435‡ | +0·163 | +0·043 |
| | 11.00 | 12.00 | 13.00 | 14.00 | 15.00 | 16.00 |
| | −0·013 | −0·106 | −0·054 | −0·075 | +0·006 | −0·057 |
| Time | 17.00 | 18.00 | 19.00 | 20.00 | 21.00 | 22.00 |
| r | −0·042 | +0·060 | −0·016 | −0·114 | −0·239† | −0·167 |
| | 23.00 | — | 01.00 | — | 03.00 | — |
| | −0·229* | — | −0·207* | — | −0·025 | — |

\* $p$ (one-tailed) < 0·05;   † $p$ (one-tailed) < 0·025;   ‡ $p$ (one-tailed) < 0·001.

Twenty-two of the subjects had a Heron Inventory score of 2 or less (extraverts), and 25 a score of 5 or more (introverts). Average temperatures for these 2 groups were computed separately at each time of day. Two-point rolling means of the resulting values are shown in Fig. 1.

It will be seen that the temperature of the "introvert" group rose more rapidly in the early morning, and started to fall at an earlier point in the late evening, than the temperature of the "extravert" group. Analysis of variance of the temperatures of the 2 groups revealed as tatistically significant interaction between Inventory score and time of day; the differences in the rhythms can therefore be taken as reliable.

*One of the objects of the experiments now to be described was to attempt to relate this new finding on introversion–extraversion to the various other separate pieces of evidence available on individual differences in rhythms of both temperature and performance, by taking both of the latter measures simultaneously at different times of day from subjects whose introversion–extraversion rating was already known. In this situation it could be predicted that, if each of the relationships already observed held true, the more introverted subjects would show rhythms in both temperature and performance that were phase advanced with respect to those of the more extraverted subjects.*

## II. The Experiments

A. *Method*

1. *Experimental design.* Because of problems of administration and of subject availability only a limited number of times of day could be chosen

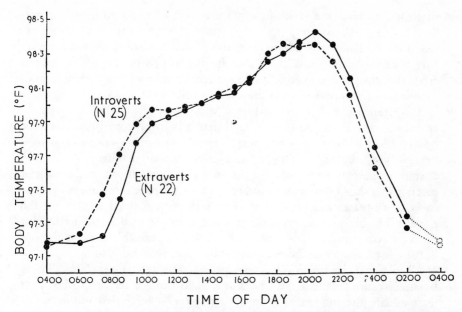

FIG. 1. Body-temperature rhythms of introverts and extraverts. Reproduced with permission from Blake (1967).

for test periods, and subjects could be tested on only a limited total number of occasions. In addition, the particular times themselves were restricted by 2 factors. Firstly, since the subjects were getting no special reward for their participation in these experiments the majority of tests had to be planned to fit in with their normal working day (08.00–16.30); and secondly, as it was considered unlikely that diurnal changes in performance would be detectable with only short intervals between test times, these times had to be spaced apart as far as was possible. In the light of these considerations, the following 5 times of day were selected: 08.00, 10.30, 13.00, 15.30, and 21.00. The first 4 of these times represent the effective limits possible within the normal working day, since the subjects had a lunch break from 11.30–13.00. The final time was chosen to coincide approximately with the predicted maximum body temperature for subjects of this type (cf. Fig. 1), in order to cover the greatest possible range of temperature variation.

Each subject was tested once at each time of day, the order in which different subjects met the different times being prescribed by balanced Latin Square designs (Fisher and Yates, 1953).

Objections could well be raised against the use of "repeated measurement" designs of this kind. Poulton and Freeman (1966) have suggested that the first trial in a series in some way sets a "standard level" of performance for the subject, who then tends to repeat this level on subsequent trials. They conclude that the use of this kind of design may therefore give misleading results, in that if one fails to get significant effects from the independent variable one cannot be sure whether there is genuinely no difference between the experimental "treatments", or whether a real difference has been masked by a "carry-over" from the first trial. In answer to this, it should only be said that it is to be hoped that any time of day effect is not of so ephemeral a nature as to be swamped out by a "carry-over" factor. A time of day effect which manifests itself as a change in performance only on the first occasion on which the subject meets the task is obviously of interest, but it must be remembered that in most work-places essentially the same task is performed all day, so the novelty has long since worn off. Indeed it might well be said that, for most practical applications, time of day effects obtainable only by the use of "single measurement" designs would be positively misleading.

All subjects were given at least one prior practice trial on each test. These practice sessions were always held at a fixed time of day, in order to at least equate the influence of any "carry-over" effects that might result from them. To avoid any specific fatigue effects, there was always an interval of at least 12 h between successive presentations of the same task. In effect this usually meant a day's interval, although sometimes trials were held at 08.00 and then 21.00 on the same day.

2. *Tasks chosen for examination.* Previous studies of the effect on performance of various kinds of "fatigue" have tended to show that it is the relatively extended uninteresting tasks that are most susceptible. Tasks of this kind were therefore the first to be chosen for investigation. Clearly, however, other kinds of task must also be examined, and, theoretically, much can be elucidated about the nature of the diurnal effect by observation of the relative levels of performance at various times of day manifested by tasks representative of different categories such as "long boring", "short complex" etc. If we take the example of loss of sleep studies (Wilkinson, 1958) we find that almost as much information about the nature of the effect of sleep-loss is to be obtained from finding that some tasks are **not** susceptible as from finding that certain tasks are. In particular, in this case it was found that tasks classified as "short complex", such as a 2-min card sorting test, showed no significant effect of 1 night's sleep deprivation, but that extended "vigilance" tasks did. It was on this evidence that it was suggested that lack of sleep manifested itself as a lowered state of arousal, this being a "reversible" process; when the task

was intrinsically arousing the sleep-deprived subject would be sufficiently stimulated to perform at normal levels.

A similar approach has been attempted here. Several different kinds of task have been studied, in order to try to elucidate which types of activity are most susceptible to any diurnal effects, and to compare these with the effects of sleep loss. If changes in performance due to time of day are, in contrast to those due to sleep loss, of an "irreversible" nature, then it would be expected that no difference in the arousal value of different tasks would alter the basic pattern of performance observed.

*In carrying out this plan, the author had completed experiments on the following eight tasks at the time of his death:*

(i) *Letter cancellation: checking through pages of English prose to cancel each letter "e" in the text.*

(ii) *Vigilance: detection of an occasional slight difference in the duration of a regularly repeated short tone.*

(iii) *Calculations: summing columns of 5 2-digit numbers.*

(iv) *Serial reaction: "chasing" a random sequence of 5 stimulus lights by tapping a corresponding response plate with a stylus.*

(v) *Card sorting: into 2 or 8 categories.*

(vi) *Paced reaction time: simple, to the extinction of a lamp.*

(vii) *Digit span: immediate recall of a sequence of numbers.*

(viii) *Time estimation: of intervals ranging from 10 to 120 sec.*

*The length of these tasks varied from a few minutes to the best part of an hour; some were "paced", others "unpaced". "Complexity" also varied somewhat, as did the degree of cognitive activity required. Although it is difficult to make a systematic classification, it is nevertheless apparent that quite a wide range of skills or abilities was represented. More details of each of the tasks are given in later sections of this chapter.*

## 3. *General procedure*

(a) The subjects. All subjects were Naval ratings with ages ranging from 17 to 33, who were sent to the laboratory in groups of 6 at a time. The majority came for a stay of 6 weeks as volunteers primarily for the long-term experiments described in the previous chapter; the remainder of the subjects came for a period of 2 weeks only. Successive groups of 6-week subjects were normally tested during the first fortnight of their stay, in order that their degree of general "test sophistication" would be comparable with the remainder. However, in one experiment where **all** the subjects were 6-weekers, the tests were carried out in the sixth week; this was done for a special purpose which will be explained later.

The subjects were housed in a small barracks where they slept and took all their meals. Since this accommodation was 1 mile from the laboratory it was not possible to control the subjects' living routine as closely as would have been desirable. They were requested to abstain from drinking alcohol during the lunch break if they were scheduled for any test in the afternoon or evening, and not to drink excessively at night at any time. It was occasionally evident that these instructions had not been complied with; in such a case the subject would be dropped from that experiment if a replacement was available.

(b) Environmental testing conditions. In order to be able to make comparisons between different tasks, the testing environment had to be held constant. Since the importance of isolating the subject had been demonstrated by Colquhoun and Corcoran in their earlier study, it was decided to maintain complete separation of each subject from his fellows in all the present experiments. This was achieved by the use of sound-proofed cubicles, the design of which was described in the previous chapter.

(c) Rating of introversion–extraversion. This was obtained from Part II (the "unsociability" scale) of the Heron Personality Inventory previously mentioned. The Inventory was administered to the subjects on the day of their arrival at the laboratory by staff unconnected with the present investigation.

(d) Measurement of body temperature. Readings were taken with a standard clinical thermometer, inserted sub-lingually for 3 min immediately before and immediately after each test session. The mean of these 2 readings was the value used in subsequent analyses.

(e) Incentives. No specific incentives were offered to the subjects for any experiment, and no knowledge of results was given at any time. Thus a subject had no information about how he was doing in any trial other than that obtainable from the task itself.

(f) Treatment and analysis of the data. The data consisted of (i) performance scores, (ii) introversion–extraversion ratings, and (iii) body-temperatures. It was dealt with in the following stages:

1. An analysis of variance was carried out on the raw scores for each performance measure obtained. From this analysis an estimate was obtained of the statistical significance (a) of the "time of day" effect (i.e., the overall differences between the mean scores at the five testing times); and (b) of the "practice" effect (i.e., the overall differences between the mean scores on successive trials).

2. Since in most cases the "practice" effect was statistically significant, it was necessary to adjust each subject's scores to allow for this effect, in order that individual "time of day" trends could be directly compared. This adjustment was based on the trial number of any particular test and

the overall practice effect for that trial for the group of subjects considered as a whole. Each subject's adjusted score at each of the 5 times of day was then converted to a percentage of his total score for all 5 trials. This procedure was necessary in order to take account of differences between subjects in overall level of performance (due, for example, to specific experience with the particular kind of task) which would not on any reasonable argument be expected to be related in any way either to time of day or to introversion–extraversion. The values obtained after these adjustments will be referred to as "adjusted" performance scores.

3. Body-temperature readings at each of the 5 testing times were converted to a percentage of the overall total in the same manner as described in Stage 2 for performance scores. This procedure was thought desirable in that, once again, it removed any individual differences due to factors (in this case physiological ones) which might be unrelated either to time of day or to introversion–extraversion. The resulting values will be referred to as "adjusted" temperatures.

4. Rank correlation coefficients (Spearman's *rho*) were computed, at each of the 5 testing times, between (a) introversion–extraversion rating and adjusted temperature, (b) introversion–extraversion rating and adjusted performance score, and (c) adjusted temperature and adjusted performance score. These correlations formed the main bases on which the relationships between circadian rhythms and personality were evaluated.

## B. *Results*

### 1. *Experiment 1: Letter Cancellation.*

(a) The task. The first experiment examined prolonged performance at a simple letter cancellation task. The letter to be cancelled was the most frequently occurring in the alphabet, "e".

Performance at this task involves virtually no other skill except that of being able to read. It is therefore usually found to be an extremely boring task, especially when, on repeated testing, the same material has always to be checked through. Output is consequently largely dependent on the subject's incentive; this may allow us to consider it analogous to some repetitive tasks in industry.

This task was chosen for the first experiment for 2 reasons. Firstly, because it is exactly the sort of test that one could predict would show the effects of lack of sleep (Wilkinson, 1965), and thus, in so far as parallels between time of day effects and those from lack of sleep exist, this task would be expected to reflect them. Secondly, it will be remembered that Colquhoun and Corcoran (1964) observed a significant positive correlation between a subject's degree of introversion and the number of lines checked in a test conducted with this task at 08.30.

Twenty-five subjects took part in this experiment. They were presented with the same material used by Colquhoun and Corcoran, namely, a cyclostyled story taken from a copy of *Punch*. The test continued for 30 min, and all subjects were given an initial practice trial of the same length on a separate day. The instructions were to "cross out as quickly and as accurately as possible all the "e"s in this piece of prose".

(b) Results. Two performance scores were obtained: (i) the total number of letters "e" processed, i.e., the total number the subject would have checked if he had not missed any (output); (ii) the percentage of letters "e" uncancelled (error).

Analysis of variance of the output scores gave a "treatment" (i.e., time of day) $F$-ratio of 12·69 ($p < 0·001$). A similar analysis of the error scores gave an $F$-ratio of 1·14 (NS). The results for output demonstrated that there was a significant variation in this aspect of performance, whereas those for error indicated that this measure was unaltered by the time of day at which the subjects were tested. The mean output scores are shown in Fig. 2, together with the corresponding mean temperature readings.

The trends in output and in body temperature through the day were closely parallel except at the 13.00 testing time, where there was a "dip" in the performance curve. The levels of both the performance and the temperature measures were lowest at 08.00 and highest at 21.00, output at the latter time being 12·6% greater than at the former.

*It is interesting to note that the "post-lunch" effect in this task appeared to be restricted almost entirely to performance at the test-time closest to the meal itself, i.e., 13.00; this suggests that it was relatively short-lived. However, to anticipate the findings of further experiments which will be described later, the post-lunch effect in other tasks sometimes appeared to continue for a considerably greater time after the meal (as indeed, it was found to do in the shift-work experiments described in the previous chapter), in that depressed performance was observed not only at 13.00 but at 15.30 also. For this reason, in much of what follows our main concern will be with the results that were obtained at the two most extreme test times, i.e., 08.00 and 21.00. Apart from being free of any post-lunch effect, these two times also in fact reflected the greatest difference in temperature observed during the "testing" day, and therefore provide perhaps the best opportunity for detecting any relationships that may exist between the variables under study.*

The correlations between introversion and adjusted temperature, introversion and adjusted performance, and adjusted temperature and adjusted performance at each of the 5 testing times are given in Table II.

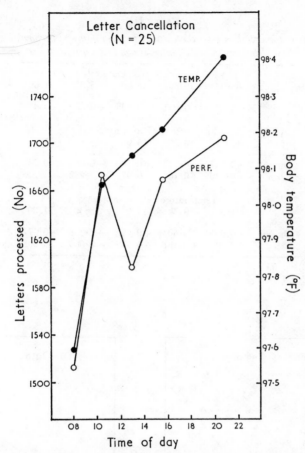

FIG. 2. Mean temperatures and mean output scores in the Letter Cancellation test (Experiment 1).

There was a significant **positive** correlation between **introversion** and **temperature** at 08.00, and a significant **negative** correlation at 21.00; the difference between these correlations was itself statistically significant ($p < 0.05$). Thus it seems reasonable to conclude that the subjects employed in this experiment were comparable to the larger sample in which a similar change in the correlation between temperature and introversion was first demonstrated. The trends in adjusted temperature during the day for the 12 "introverts" (taken as those subjects with Heron scores of 5 or more) and the 10 "extraverts" (taken as those subjects with Heron scores of 3 or less) are shown in Fig. 3a, in which the difference in the phasing of the rhythm in the 2 groups can be seen to correspond approximately with that found in the larger sample (cf. Fig. 1).

E

TABLE II

*Letter Cancellation test: correlations between adjusted performance,
adjusted temperature, and introversion at each time of testing*

|  | Introversion × Temperature | | | | |
|---|---|---|---|---|---|
|  | 08.00 | 10.30 | 13.00 | 15.30 | 21.00 |
|  | +0·382* | −0·051 | −0·105 | +0·229 | −0·430* |

|  | Introversion × Performance | | | | |
|---|---|---|---|---|---|
|  | 08.00 | 10.30 | 13.00 | 15.30 | 21.00 |
| Output | +0·417* | −0·170 | +0·180 | −0·079 | −0·283 |
| Error | −0·240 | +0·168 | +0·077 | −0·252 | +0·270 |

|  | Temperature × Performance | | | | |
|---|---|---|---|---|---|
|  | 08.00 | 10.30 | 13.00 | 15.30 | 21.00 |
| Output | +0·345* | +0·603† | +0·192 | +0·338* | +0·477† |
| Error | +0·033 | +0·151 | −0·139 | −0·065 | −0·005 |

\* $p$ (one-tailed) $< 0·05$
† $p$ (one-tailed) $< 0·01$

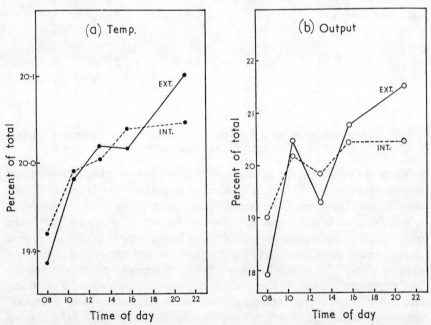

FIG. 3. Mean adjusted temperatures (a) and mean adjusted output scores
(b) for introverts and extraverts in the Letter Cancellation test.

3. TEMPERAMENT AND TIME OF DAY

As predicted by the results of the experiment by Colquhoun and Corcoran, the correlation between **output** and **introversion** was positive (and statistically significant) at 08.00, whereas error was not significantly related to personality at this (or any other) testing time. The correlations between output and introversion at testing times other than 08.00 were not statistically significant, but there was a relatively large **negative** correlation at 21.00, which was significantly different from the correlation at 08.00 ($p < 0.05$). The trends in adjusted output during the day for introverts and extraverts are shown in Fig. 3b. Comparison of these trends with those in Fig. 3a shows that there was a close correspondence between temperature and performance rhythms in each group, except at the 13.00 testing time, where post-lunch depression of performance was evident in both introverts and extraverts. For the introvert group, a mean change in adjusted temperature of 0·13 percentage points between 08.00 and 21.00 was accompanied by a mean change in adjusted output of 1·44 percentage points; the corresponding figures for the extravert group were 0·22 and 3·57 respectively.

There was a clear positive correlation between **temperature** and **output** at all 5 testing times; it was only at 13.00 that the correlation was not statistically significant. By combining the probabilities associated with each of the individual coefficients on the assumption of independence (Fisher's method) it was possible to compute the probability that all 5 correlations would occur together by chance alone. This was found to be less than one in a thousand.

*These last correlations are of great interest, since, as far as is known, this is the first time that a relationship between the level of "normal" body temperature and an index of performance efficiency has been demonstrated where fluctuations in either measure due to time of day have been, as it were, "held constant". In previous experiments where temperature and efficiency were found to be related (e.g., Kleitman, Titelbaum and Fievson, 1938) the possibly confounding effect of time of day as a "third factor" which is itself correlated with both temperature and performance was not controlled, since the results were based on data obtained from single subjects at different times, rather than, as here, from a group of subjects at the same time.*

*2. Experiments 2–6. The following 5 experiments produced results which were, in most respects, essentially similar to those found with the Letter Cancellation test. However, some of the findings were not as clear cut as they were in that case. In order to make the presentation more coherent, therefore, the results from these 5 experiments have been grouped together, in the hope that any general patterns that exist will be seen more readily.*

(a) The tasks.

(i) Experiment 2: Vigilance. This task, an adaptation of one devised by Wilkinson (1964), was, in effect, an auditory analogy of the Clock Test (Mackworth, 1948). It consisted of a sequence of tones each lasting (normally) 600 msec, and occurring every 3 sec throughout the test session, which was of 53 min duration. Randomly distributed within the sequence of tones were 24 tones of slightly longer duration (670 msec); these were the "signals" to be detected. Records were kept of correct responses to these signals, and also of any "false reports" made. Since the discrimination required was a difficult one, 5 separate 1 h practice trials were given before the actual test sessions were held. Twenty-five subjects were tested in this experiment.

(ii) Experiment 3: Calculations. This was the same task used in the shift-work experiments described in the previous chapter. The length of the test session was 1 h. By deliberate choice, all the 25 subjects used in this experiment were "6-week" subjects who had already completed a 12-day shift-work trial; they were thus **highly practised in the task**. As before, they were asked to work as rapidly and accurately as possible. The two scores obtained were (1) the number of sums attempted (output), and (2) the percentage of incorrect answers (error).

(iii) Experiment 4: Serial Reaction. Leonard's 5-choice test of serial reaction (Leonard, 1959) is a task which has frequently shown itself to be sensitive to many types of "fatigue" or stress (see Table I in Poulton, 1966). The subject is confronted with a display of 5 small lamps mounted in a pentagon formation. On a table in front of this display is a board with 5 metal discs which are arranged in a pattern that matches that of the lights. One of the lamps is lit at the start of the test and the subject has to tap the appropriate disc with a stylus to put it out; this action automatically lights another lamp and the subject again has to tap the appropriate disc, and so on. The order in which the lamps are lit is random. The subject is told to carry out the task "as quickly and accurately as possible". Scores are taken of correct responses; errors, i.e., when the subject taps an inappropriate disc; and "gaps", one of which is recorded whenever the subject allows an interval of 1·5 sec or more to elapse between successive taps of the stylus. Each test in the present series lasted 30 min. Thirty subjects took part in this experiment. They were given a single practice trial before the test sessions proper were held.

(iv) Experiment 5: Card Sorting. Cards were sorted into 2 categories and 8 categories. Packs of 64 cards were made up from 2 decks of ordinary playing cards with the 9, 10, and court cards removed. The 2-category task was to divide 4 of these packs by colour only; the 8-category task was to sort a further 4 packs by face value. The whole test occupied some 12 to 15 min, and was always preceded by a short practice run. Subjects were instructed to sort the cards as quickly and accurately as possible, but to ignore any mis-sorts they noticed. The scores were the mean times to sort the 4 packs in each category. Thirty subjects were employed.

(v) Experiment 6: Paced Reaction Time. The task here was "simple" reaction time to the extinction of a bright lamp. This lamp was switched off at approximately 15-sec intervals, with fore-periods (illumination time) varying between 3 and 5 sec. The test lasted for some 20 min. Subjects were asked to depress a push-button switch as quickly as possible when the light went out. Response times were recorded to an accuracy of 0·01 sec by "Dekatron" timers, and the score taken was the mean of the 75 $RT$s obtained. Twenty-five subjects were tested in this experiment; they were given a single practice trial before the test series was held.

(b) Results.

(i) Overall time of day effects. The $F$-ratios for the "time of day" effect derived from the analyses of variance carried out on each of the scores from the 5 tests are given in Table III, together with the level of statistical significance associated with them.

TABLE III

*Experiments 2–6:* F-ratios derived from analyses of variance of the performance scores

| Test | Score | F-ratio | p (two-tailed) |
|---|---|---|---|
| Vigilance | Detections | 3·55 | <0·025 |
|  | False reports | 0·98 | (NS) |
| Calculations | Output | 5·65 | <0·001 |
|  | Error | 0·76 | (NS) |
| Serial Reaction | Correct responses | 2·78 | <0·05 |
|  | Gaps | 2·56 | <0·05 |
|  | Errors | 1·54 | (NS) |
| Card Sorting | 2-category time | } 7·14* | <0·001 |
|  | 8-category time |  |  |
| Reaction Time | Mean $RT$ | 1·81 | (NS) |

* Based on the combined scores for both categories; there was no significant interaction between time of day and category size.

Significant variation through the day was observed for the following scores: Vigilance detections; Calculations output: Serial Reaction correct responses, and also gaps; and Card Sorting time (both 2-category and 8-category). The $F$-ratio for mean $RT$ in the Reaction Time test was not significant, but the difference between the scores at 08.00 and 21.00 was found to be significant at the 5% level of confidence by a Wilcoxon matched-pairs signed-ranks test (Siegel, 1956) on the individual subjects' adjusted times. No significant variation was observed for Vigilance false reports, Calculations error, and Serial Reaction errors; these scores have therefore been omitted from Fig. 4, in which the mean values of the various performance measures at the 5 testing times are shown for each test, together with the mean temperature readings recorded (note that the scales for Serial Reaction gaps, Card Sorting times, and Reaction Time means have been inverted).

Not all the temperature curves showed the smooth shape obtained from the larger sample (cf. Fig. 1); this was perhaps only to be expected. Nevertheless, in each case the mean value was considerably higher at 21.00 than at 08.00, though, again, the extent of the observed difference varied somewhat, and was in all cases smaller than recorded from the larger sample.

A clear improvement in efficiency between the first and last testing time was evident in all performance scores. The trends for the mean scores at the intervening test times appeared in all cases to be disturbed by post-lunch effects at 13.00 or 15.30 or both. Nevertheless, each performance curve was characterized by improvement between 15.30 and 21.00. In addition, the performance level at 10.30 was higher than that at 08.00 in all scores except Serial Reaction correct responses. Thus it seems reasonable to conclude that the basic "time of day" effect observed in the output score of the Letter Cancellation test was also present in each of the performance scores shown in Fig. 4. Note that this group includes the output score from the Calculations test. This is important, since it demonstrates that clear time of day effects can be found even when the task is one which has been extensively and specifically practised.

Altogether then, 6 tasks (8 scores) have so far been shown to be sensitive to time of day effects. The improvement in performance between 08.00 and 21.00 for these 8 scores, expressed as a percentage change, is shown in Table IV, together with the corresponding increases in temperature. The ratio of these values is also given; this ratio allows for variation in the extent of temperature increase in the different subject groups, and thus perhaps provides a more valid basis for comparing the extent of the time of day effects in the different tasks.

The two scores in which the greatest improvement was found between 08.00 and 21.00 were Vigilance detections and Serial Reaction gaps. As it happens, the 2 tests from which these scores were obtained were the only ones in which performance measures were available for sub-sections of the test session as well as for the entire working period. Thus an additional examination of the time of day effect on the scores from these 2 tests can be made by plotting performance levels as a function of time on task. Scores for the first and the second halves of each test session are shown in

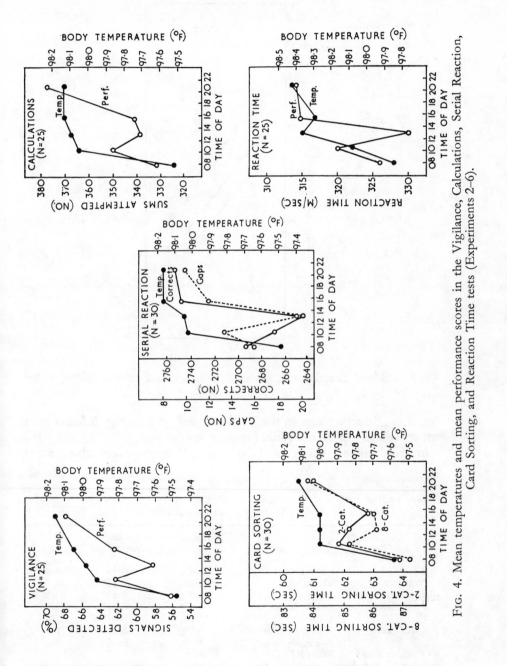

Fig. 4. Mean temperatures and mean performance scores in the Vigilance, Calculations, Serial Reaction, Card Sorting, and Reaction Time tests (Experiments 2–6).

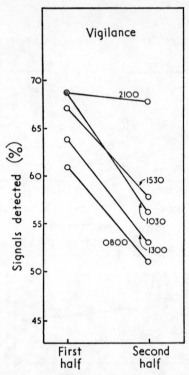

Fig. 5. Mean detection-rate scores in the first and second halves of the
Vigilance test.

Fig. 5 for detection rate in the Vigilance test, and in Fig. 6 for correct
responses and gaps in the Serial Reaction test.

In the Vigilance test typical "decrement" curves were observed at

TABLE IV

*Percent improvement in performance, and absolute increase in temperature, between
08.00 and 21.00 for 8 scores from Experiments 1–6*

| Test | Score | Performance Improvement (A) | Temp Increase (B) | Ratio: (A)/(B) |
|---|---|---|---|---|
| Letter Cancellation | Output | 12·6% | 0·82° | 15·4 |
| Vigilance | Detections | 21·4% | 0·68° | 31·5 |
| Calculations | Output | 13·9% | 0·61° | 22·8 |
| Serial Reaction | Correct responses | 2·2% | 0·67° | 3·3 |
| | Gaps | 37·1% | 0·67° | 55·4 |
| Card Sorting | 2-category time | 4·7% | 0·55° | 8·6 |
| | 8-category time | 3·4% | 0·55° | 6·6 |
| Reaction Time | Mean *RT* | 3·7% | 0·57° | 6·5 |

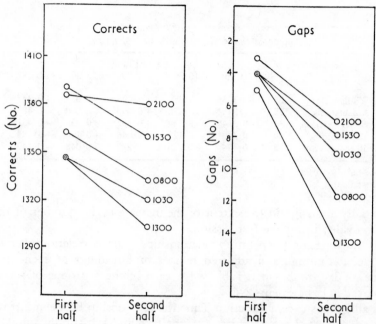

Fig. 6. Mean performance scores in the first and second halves of the Serial Reaction test.

each test time except 21.00, where detection rate remained almost unaltered during the session. At each of the remaining times the fall-off in performance was substantial, but of essentially the same magnitude. Thus in respect of the difference in score between 08.00 and 21.00, the "time of day" effect was clearly greater in the second half of this test than in the first.

In the Serial Reaction test there was again evidence of within-session decrement. This occurred both in the correct responses score and in the gaps score. In the case of correct responses the picture was similar to that for detection rate in the Vigilance test, i.e., negligible decrement at 21.00, and larger, but similarly sized decrements at the remaining times. However, the relatively small magnitude of the decrements meant that the difference in the 08.00–21.00 time of day effect in the first and second halves of the session was not as marked as in the Vigilance case.

The gaps score results showed clearly that the time of day effect on this aspect of performance was confined almost entirely to the second half of the test; within-session decrement was substantial, and perfectly rank-correlated with overall score. Another way of describing the findings here would be to say that the time of day effect on gaps was in fact

TABLE V

*Experiments 2–6: correlations between introversion and
adjusted temperature at each time of testing*

| Experiment | N | Test time | | | | |
| | | 08.00 | 10.30 | 13.00 | 15.30 | 21.00 |
|---|---|---|---|---|---|---|
| Vigilance | 25 | +0·365* | +0·255 | +0·096 | −0·072 | −0·337* |
| Calculations | 25 | +0·345* | +0·176 | +0·098 | +0·090 | −0·487† |
| Serial Reaction | 30 | +0·364* | −0·339* | +0·332* | −0·135 | −0·451† |
| Card Sorting | 30 | +0·340* | −0·121 | −0·046 | +0·148 | −0·188 |
| Reaction Time | 25 | −0·174 | +0·226 | +0·222 | +0·048 | −0·149 |

\* $p$ (one-tailed) < 0·05
† $p$ (one-tailed) < 0·01

essentially a change in the extent of the increase in the number of these lapses made during the test session.

(ii) Introversion-temperature relationships. The correlations between introversion ratings and adjusted temperature readings at each of the 5 times of day are shown in Table V for each of the 5 groups of subjects employed in Experiments 2–6.

In all groups except Reaction Time the relationship between introversion and adjusted temperature was **positive** and statistically significant at 08.00. At 21.00 the relationship was **negative** in all groups, but statistically significant only for the Vigilance, Calculations and Serial Reaction cases. The **difference** between the correlations at 08.00 and 21.00 was statistically significant for Vigilance, Calculations, Serial Reaction and Card Sorting (though only marginally for the last-named group). At the intervening test times the only significant correlations observed were in the Serial Reaction group at 10.30 (negative) and 13.00 (positive).

In the results for the larger sample described earlier, a significant positive correlation was found at 08.00, and a significant negative correlation at 21.00. None of the correlations at intermediate times was statistically significant. Thus on the basis of the results given in Table V we may state with a reasonable degree of certainty that both the Vigilance and the Calculations groups were essentially similar in make-up to the larger sample (as was the Letter Cancellation group). We can be slightly less sure of this point in the Serial Reaction, Card Sorting, and Reaction Time groups; however, further information on the comparability of these subject samples can be obtained from the curves of adjusted temperature computed separately for extraverts and introverts in each case (see Fig. 7).

In all groups the temperature of the extraverted subjects was relatively

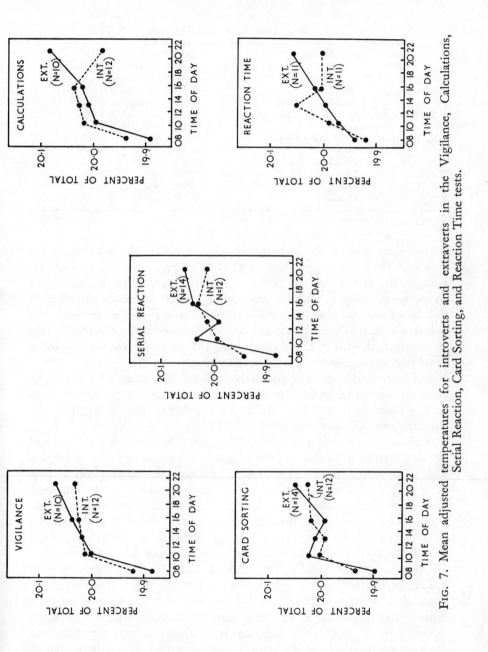

Fig. 7. Mean adjusted temperatures for introverts and extraverts in the Vigilance, Calculations, Serial Reaction, Card Sorting, and Reaction Time tests.

TABLE VI

*Experiments 2–6: correlations between introversion and adjusted performance at each time of testing*

| Test | Score | Test Time | | | | |
|------|-------|-----------|---|---|---|---|
| | | 08.00 | 10.30 | 13.00 | 15.30 | 21.00 |
| Vigilance | Detections | +0·144 | +0·037 | −0·072 | +0·053 | −0·219 |
| | False Reports | −0·246 | +0·221 | +0·319 | −0·065 | −0·137 |
| Calculations | Output | +0·263 | −0·138 | −0·171 | +0·188 | −0·152 |
| | Error | −0·078 | −0·125 | +0·075 | +0·148 | −0·038 |
| Serial Reaction | Correct responses | −0·184 | +0·175 | −0·080 | −0·001 | +0·124 |
| | Gaps | −0·106 | +0·089 | −0·255 | +0·275 | −0·017 |
| | Errors | +0·044 | −0·028 | −0·156 | +0·201 | +0·048 |
| Card Sorting | 2-category time | +0·122 | +0·079 | −0·546* | +0·134 | +0·183 |
| | 8-category time | +0·276 | +0·061 | −0·436* | +0·228 | +0·004 |
| Reaction Time | Mean $RT$ | +0·156 | +0·129 | −0·156 | +0·091 | +0·126 |

\* $p$ (one-tailed) $< 0·01$

higher than that of the introverted subjects at 21.00, and (except for Reaction Time) the reverse was true at 08.00. Thus, apart from 1 discrepant finding, the end-points of the curves indicate that the phase advance in the temperature rhythm of introverted subjects originally observed in the larger sample also occurred in the present 6 smaller groups. The subjects in each of these experiments will therefore be taken as comparable in this respect both to those comprising the larger sample, and to those employed in the Letter Cancellation experiment.

(iii) Introversion–performance relationships. The correlations between introversion ratings and adjusted performance scores at each of the 5 times of day are shown in Table VI for the 5 different tasks comprising the present group of experiments.

The only statistically significant correlations observed were in the Card Sorting test; in both the 2-category and the 8-category versions of this task there was a negative correlation between introversion and sorting time at 13.00. However, it could be argued that these 2 correlations were not strictly independent of each other. Essentially then we have only 1 significant coefficient in the entire Table. The probability of this occurring by chance is itself high enough to lead us to suspect the validity of the correlation in question. We must therefore conclude that no relationship between introversion and performance has been definitely shown to exist for any of the tasks in this group. However, it is worth noting that in the case of 2 scores—Vigilance detections, and Calculations output—the sign of the observed correlation changed from positive to negative between 08.00 and 21.00; although in neither case was the difference between the correlations statistically significant, the direction of change

TABLE VII

*Experiments 2–6: correlations between adjusted temperature and adjusted performance at each time of testing*

| Test | Score | Test Time | | | | |
|---|---|---|---|---|---|---|
| | | 08.00 | 10.30 | 13.00 | 15.30 | 21.00 |
| Vigilance | Detections | +0·231 | −0·203 | +0·097 | +0·111 | −0·056 |
| | False Reports | −0·397* | −0·162 | +0·136 | +0·179 | −0·012 |
| Calculations | Output | +0·235 | +0·014 | +0·121 | +0·258 | +0·015 |
| | Error | −0·053 | +0·086 | −0·252 | −0·224 | −0·075 |
| Serial Reaction | Correct responses | +0·148 | +0·124 | +0·167 | +0·339* | +0·061 |
| | Gaps | −0·131 | −0·022 | −0·028 | −0·009 | +0·127 |
| | Errors | +0·063 | +0·250 | −0·025 | +0·307* | −0·112 |
| Card Sorting | 2-category time | −0·083 | +0·150 | +0·213 | −0·330* | −0·097 |
| | 8-category time | −0·213 | −0·091 | −0·071 | −0·259 | −0·332* |
| Reaction Time | Mean $RT$ | −0·112 | −0·070 | −0·367* | +0·075 | −0·019 |

* $p$ (one-tailed) $< 0·05$

was in fact the same as that found in the Letter Cancellation test, where it was proved reliable. It would therefore seem worthwhile to include tasks of these 2 kinds in any further investigation of the introversion–performance relationship that may be undertaken.

(iv) Temperature–performance relationships. The correlations between adjusted temperatures and adjusted performance scores at each of the 5 times of day are shown in Table VII.

Since there is no way in which one could predict the manner in which any relationship between temperature and performance would be affected by the time of day at which the measures were taken, we must assume for the present that there is no such effect, and look simply for consistency in the direction of any such correlation that may be present in a particular task. In general, we would presumably expect to find higher temperatures associated with better performance rather than the reverse, and, indeed, such a positive association appeared quite reliably in the case of the output score in the Letter Cancellation experiment described earlier. In the present group of experiments coefficients of the same sign at all 5 test times occurred in the results for Calculations output, Serial Reaction correct responses and Card Sorting (eight-category time). In addition, coefficients of the same sign at 4 of the 5 test times were present in the results for Calculations error, Serial Reaction gaps, and Reaction Time mean $RT$. In all cases the direction of the relationship appeared to be such that higher temperatures were, in fact, related to better performance; however, combination of the probabilities associated with each of the 5 observed correlation coefficients for each score (on the assumption of independence) indicated that the relationship was statistically significant at an acceptable level of confidence only for Calculations output ($p < 0·05$), Serial Reaction correct responses ($p < 0.01$), and Card Sorting ($p < 0.01$). The reliability of the results for Reaction Time mean $RT$ was marginal ($p = 0.10$); the remaining cases were not acceptable ($p > 0.10$).

*It is of interest to note that each of the 4 scores in which a relationship appeared to be present was, as was Letter Cancellation output, a measure of speed (since error rates in the Serial Reaction task were very low, the score of correct responses is essentially one of output). The implication of this finding is discussed later.*

3. *Experiment 7: Digit Span. In all the experiments described so far (with the possible exception of Experiment 2—Calculations) the task employed was a relatively simple one, not requiring any great degree of cognitive or intellectual effort. Efficiency at the task used in the present experiment is usually held to be related to general intelligence, and thus could be said to represent a distinctly higher type of mental activity.*

(a) The task. The task was the "digit-span" test as used in the well-known Stanford–Binet Intelligence Scale (Terman and Merrill, 1937). Single digits were read out to the subject at the rate of one per second, in sequences which progressively increased in length as the test proceeded. The order of digits in the sequences was at random. The test commenced with sequences containing 3 digits only. If the subject was able to repeat each of 3 such sequences without error, a further set of 3 sequences each containing 4 digits was presented. The test was terminated at the point where the subject was unable to repeat the complete set of sequences of a given length. Scoring was as follows: if a subject successfully recalled all 3 of the set of sequences of length (say) 7 digits, but none of the set of length 8, his span would be assessed at 7·0 digits. If he recalled all 3 of length 8 but none of length 9 his span would be 8·0, and so on. If a subject recalled only 1, or 2, of the sequences in a set of a given length without error a proportionate score was assigned, e.g., all 3 sequences of length 7, and one of length 8 recalled, score: 7·33 digits. Thirty subjects were used in this experiment; the test was typically completed within some 5 min of commencement.

(b) Results. Analysis of variance of the digit-span scores indicated that there was a significant variation in efficiency at this test at the different testing times ($F = 3.41$, $p < 0.05$). The mean scores at the 5 times are shown in Fig. 8, together with the corresponding mean temperatures.

The trend in mean performance levels through the day departed clearly from that observed in all previous cases where a significant time of day effect was observed; in those cases the increase in body temperature during the day was consistently reflected in an overall **improvement** in performance. The present results show that an initial increase in digit span between 08.00 and 10.30 was followed by a steady **decline** in the level achieved

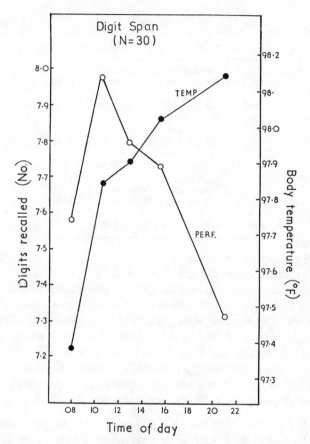

FIG. 8. Mean temperatures and mean performance scores in the Digit Span test (Experiment 7).

right up to the 21.00 testing time, where the mean number of digits recalled was 8·6% fewer than at 10.30. By contrast, mean temperature continued to rise during the whole "test" day.

None of the correlations between introversion, adjusted temperature and adjusted performance was statistically significant (see Table VIII). However, it is interesting to note that the **sign pattern** of the correlation between introversion and performance through the day reflected exactly the pattern observed for introversion and temperature.

The fact that the correlations between temperature and performance at the 5 testing times were small and variable is perhaps only what might be expected in view of the shape of the time of day trend. If, as this trend

TABLE VIII

*Digit Span test: correlations between adjusted performance, adjusted temperature and introversion at each time of testing*

| | Introversion × Temperature | | | |
|---|---|---|---|---|
| 08.00 | 10.30 | 13.00 | 15.30 | 21.00 |
| +0·072 | −0·254 | +0·181 | +0·209 | −0·226 |
| | Introversion × Performance | | | |
| 08.00 | 10.30 | 13.00 | 15.30 | 21.00 |
| +0·134 | −0·031 | +0·084 | +0·304 | −0·220 |
| | Temperature × Performance | | | |
| 08.00 | 10.30 | 13.00 | 15.30 | 21.00 |
| +0·155 | −0·001 | −0·062 | −0·078 | +0·050 |

suggests, the effect of increasing temperature on digit-span is non-linear, simple rank correlation would not show this.

*It is difficult to suggest just why the time of day effect in this test took the particular form observed, with performance generally deteriorating rather than improving as the day progressed. This trend could be interpreted either in terms of a negative correlation between level of arousal and short-term memory (which has been demonstrated by Kleinsmith and Kaplan, 1963), or as a direct effect of body temperature on rate of forgetting. French (1942) studied the retention of a maze habit by goldfish and noticed that retention improved as temperature was lowered; however, there do not appear to be any comparable data for man.*

4. *Experiment 8: Time Estimation. Hoagland (1933) suggested that the perception of time depends directly on a temperature-sensitive internal "clock"; the higher the temperature, the faster this clock operates. Pfaff (1968), by observing average time estimation performance at various points in the normal cycle of circadian temperature, was able to confirm the predictions of Hoagland's theory quite satisfactorily. However Pfaff's data contradicted the results of Thor (1962), which implied that the "internal clock" is running* **slower** *at high points in the daily temperature rhythm.*

*The present experiment was in effect partly an attempt to reconcile these discrepant findings, and to relate any differences found between individuals to their introversion rating and to the expected accompanying phase differences in temperature rhythm. It should be noted here that Bell and Watts (1966), in a large scale investigation, found that the number of significant associations of time estimation with personality did not differ itself significantly from chance; there was thus no basis for predicting the direction of the relationship (if any) between introversion and time estimation in the present case.*

(a) The task. Thirty subjects took part in this experiment. On each occasion of testing they were required to estimate intervals of 10, 20, 30, 60 and 120 sec by the method of production. Each interval was estimated twice, the mean time estimated on the 2 trials being taken as the score for subsequent analysis. The order in which the various intervals were estimated was permuted on successive trials. The interval to be assessed was initiated by the extinction of a lamp; when the subject estimated that the required interval had passed he depressed a switch which stopped a timer. The whole series of estimations was completed in some 20 min.

(b) Results. Separate analyses of variance of the produced times for each of the 5 intervals failed to reveal a significant overall time of day effect for any interval. The mean estimations over all 5 trials were quite accurate, but there were, as might be expected, considerable individual differences. That these were related to introversion is suggested by the mean estimations for "extraverts" ($N=13$) and "introverts" ($N=11$); these (expressed as a percentage of the interval to be estimated) were, for the 5 intervals 10, 20, 30, 60 and 120 sec 108·4%, 105·0%, 105·1%, 104·5% and 109·8% respectively (extraverts); and 99·7%, 99·4%, 99·5%. 97·4% and 96·2% (introverts). Thus the extraverts were consistently over-estimating, and the introverts under-estimating, the required interval; however, the extent of the error was relatively small.

None of the correlations between **introversion** and **adjusted temperature** was statistically significant (see Table IX). This indicates that the subjects used in the present experiment were not altogether comparable with those used in the large scale study.

Correlations between **introversion** and **adjusted performance** were also not significant, although there was some evidence of a shift from mainly positive to mainly negative coefficients between the first 3 and the last 2 test times. Examination of the trends in mean adjusted scores for the introvert and extravert groups suggested that whereas there was no evidence of any significant variation in the introverts, the **direction** of the time of day effect in the extraverts appeared to change as the requested interval was lengthened. At the 2 shortest intervals the time produced by this group was **shortest** at 08.00, and **longest** at 21.00; the mean difference between these 2 times was statistically significant ($p=0·023$, Wilcoxon test). By contrast, at the 120-sec interval the produced time was **longest** at 08.00, and second **shortest** at 21.00. This difference between the 08.00–21.00 trends at the longest and at the two shortest intervals was statistically significant ($p=0·047$), i.e., the transformation of an overall **increase** in produced time through the day to an overall **decrease**, as the interval to be estimated was increased by a factor of 6, appeared to be a genuine phenomenon in the present group of extraverted subjects.

TABLE IX

*Time Estimation test: correlations between adjusted performance,
adjusted temperature and introversion at each time of testing*

|  | Introversion × Temperature | | | | |
|---|---|---|---|---|---|
|  | 08.00 | 10.30 | 13.00 | 15.30 | 21.00 |
|  | +0·154 | −0·155 | +0·095 | −0·130 | +0·045 |

|  | Introversion × Performance | | | | |
|---|---|---|---|---|---|
|  | 08.00 | 10.30 | 13.00 | 15.30 | 21.00 |
| 10 sec | +0·120 | +0·009 | +0·224 | −0·029 | −0·150 |
| 20 sec | +0·066 | +0·090 | +0·196 | −0·052 | −0·127 |
| 30 sec | +0·009 | +0·055 | +0·276 | −0·008 | +0·026 |
| 60 sec | −0·054 | +0·276 | +0·200 | −0·302 | −0·121 |
| 120 sec | −0·107 | +0·062 | +0·063 | +0·228 | −0·043 |

|  | Temperature × Performance | | | | |
|---|---|---|---|---|---|
|  | 08.00 | 10.30 | 13.00 | 15.30 | 21.00 |
| 10 sec | −0·146 | −0·168 | −0·117 | +0·182 | −0·306* |
| 20 sec | −0·022 | −0·261 | −0·222 | +0·136 | −0·115 |
| 30 sec | −0·059 | −0·312* | −0·261 | −0·015 | +0·215 |
| 60 sec | −0·087 | −0·455† | +0·129 | +0·280 | +0·116 |
| 120 sec | −0·334* | −0·275 | −0·150 | +0·148 | +0·079 |

\* $p$ (one-tailed) $< 0·05$
† $p$ (one-tailed) $< 0·01$

The correlations between **temperature** and **adjusted performance** at
each of the 5 testing times were negative in all but 1 interval case at 08.00,
10.30 and 13.00, and positive in 7 of the 10 cases at 15.30 and 21.00.
Only certain of these correlations were statistically significant, but the
patterning was sufficiently striking to suggest that if there is indeed any
underlying relationship between temperature and time estimation then
this relationship itself may change as a function of time of day.

*The conclusions from these findings seem to be (i) that in research in this area
it is important that the introversion rating of the subjects be assessed, and (ii) that
it may be misleading to combine estimations of widely different intervals as Pfaff
(1968) did, since such an averaging process may conceal important differences in
the underlying relationship between temperature and performance in certain subjects
for estimates of varying required magnitude. Until further experiments have been
carried out with these considerations in mind it would appear that we cannot yet
state with certainty that the "internal clock" is speeded up by temperature increases
when these result solely from natural circadian variation.*

# III. Discussion

*The tasks used in the present study were designed to cover the main parameters known to influence the extent to which performance is affected by loss of sleep (Wilkinson, 1965; Corcoran, 1962a). Thus they varied in "simplicity–complexity", in "interest value", in the degree to which they resembled every-day activities such as reading (this will be referred to as "degree of practice"), in duration, and in "signal frequency" (which indicates the rate at which critical events occurred). In addition 3 different measures of performance have been sampled, each of which is known to be affected in a different manner by periods without sleep. These are*

(i) *speed of execution of the task (which we shall sometimes refer to as "output"),*

(ii) *number of omissive errors (which are failures to execute a response), and*

(iii) *number of commissive errors (which are responses made when they should not have been).*

*In general, tasks which can be described as simple, uninteresting, well practised, long, and of low signal frequency are more seriously affected by loss of sleep than tasks with some or all of the opposite characteristics. It is also well known that speed and errors of omission are affected by sleep deprivation, but that commissive errors are only affected in experimenter-paced tasks (Williams, Lubin and Goodnow 1959; Corcoran, 1962a).*

*Two other factors will be introduced in this discussion. Wilkinson (1961) showed that the depression of performance by loss of sleep can be considerably reduced by feeding the subjects information about how well they are doing; this will be known as the "knowledge of results" (KR) effect. Corcoran (1962b) and Wilkinson (1963) showed that the introduction of high levels of noise into the environment had an effect which was very similar to that obtained with KR. Later in this section 2 experiments which investigated the effects of KR and of noise on performance at different times of day will be briefly reported.*

*Thus the overall hypothesis we shall test is whether those task and environmental parameters known to affect the performance of subjects deprived of sleep are also those which affect performance at different times of day. If this is found to be the case we can deduce that the causes of performance changes at different times of day are the same as those which affect performance after loss of sleep.*

*The most economical method of organizing the data from the present experiments is in terms of a theory proposed by Corcoran (1962a) to account for the effects of loss of sleep on performance, and also for the differences in behaviour of introverted and extraverted subjects. The great majority of the present findings can be explained by Corcoran's theory, and the points at which theory and fact diverge are of interest in suggesting modifications to it.*

*In general, Corcoran's theory held that a physiological state of "arousal" or "activation" was the sole underlying variable affecting performance after loss of sleep, and that the relationship between performance and level of arousal was an inverted-U shaped function. It was further postulated that sleepiness and arousal*

TABLE X

*Characteristics of the tasks used in the experiments*

| | Exp. 1 Letter Cancellation | Exp. 2 Vigilance | Exp. 3 Calculation | Exp. 4 Serial Reaction | Exp. 5 Card Sorting 2-cat. | Exp. 5 Card Sorting 8-cat. | Exp. 6 Reaction Time | Exp. 7 Digit Span | Exp. 8 Time Estimation |
|---|---|---|---|---|---|---|---|---|---|
| Simplicity | + | + | ? | + | + | ? | + | − | − |
| Low interest | + | + | + | + | + | ? | + | ? | ? |
| Well practised | + | +* | + | − | ? | ? | ? | − | − |
| Relatively long | | + | − | + | − | − | + | − | − |
| Low signal frequency | − | + | − | − | − | − | − | − | ? |

* In general, people are well practised at "monitoring" activities.

A plus sign in the appropriate cell indicates that the task is considered to have the characteristic described in the left-hand column. If the time of day effect is similar to the effect of sleep deprivation then those tasks with the greatest number of plus signs should show the clearest time of day changes.

*were perfectly related, and therefore that deprivation of sleep had the effect of reducing level of arousal. Other factors were also considered to affect arousal level; those task parameters which increased the level of arousal reduced the effect of loss of sleep, and those which reduced the level of arousal increased the effect of loss of sleep. It was further stated that certain tasks and certain measures of performance had different optimal levels of arousal, such that the level of arousal optimal for, say, output was too high for error-free performance.*

*Having noted a number of similarities between the performance of extraverts on the one hand and of subjects deprived of sleep on the other, Corcoran hypothesized that extraverts were characterized by lower levels of arousal than introverts. The experiments designed to test this hypothesis were in general successful (e.g., Corcoran, 1964a, 1964b), and subsequent work (e.g., Davies and Hockey, 1966) has provided further support for the arousal theory of introversion-extraversion.*

*The first stage in considering the present data, then, must be to examine the similarity of the tasks that showed performance changes at different times of day to those susceptible to loss of sleep. Table X indicates the extent to which each of the tasks used in the experiments possessed those characteristics known to be conducive to sleep-loss effects.*

*It is evident that 2 of the tasks (Time Estimation and Digit Span) lack most of the characteristics normally associated with tasks susceptible to loss of sleep. In the present study Time Estimation did not show any clear time of day effects, whilst Digit Span exhibited a performance trend through the day which was actually the* **reverse** *of that shown by the remaining 6 tasks. Of these, Letter Cancellation, Vigilance, Calculations, Serial Reaction, and Paced* RT *form a similar group, and we saw in the results that in each case performance improved between morning and evening. Card Sorting is difficult to classify, thus its column contains more question marks than the others; performance at this task, however, also improved through the day.*

*To the similarity between the kinds of task susceptible to loss of sleep and to time of day indicated by Table X may be added the evidence concerning the particular measures of performance which showed the most clear time of day effects in the present experiments. If there is indeed a correspondence between the effects of loss of sleep and of time of day, then scores of* **speed** *and* **omissive** *errors should show greater diurnal variation than scores of* **commissive** *errors. In fact, in none of the tasks did commissive error rates show meaningful time of day trends, whereas measures of both speed and of omissive errors (the latter being represented by Vigilance misses and Serial Reaction gaps)* **were** *significantly related to the time of testing.*

*All these findings thus support the hypothesis that the factors underlying the time of day effect are the same as those underlying performance after deprivation of sleep, and, therefore, that* **level of arousal** *changes with time of day.*

*This theory may now be tested by deliberately introducing "arousing" factors into*

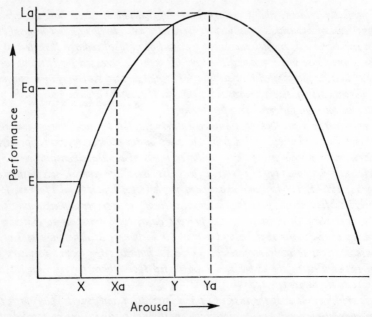

Fɪɢ. 9. Postulated relationship between level of arousal and performance efficiency.

*a task which under normal conditions shows definite time of day effects. First we must be clear about the predictions which the arousal theory as stated by Corcoran would make in this situation. Figure 9 shows the postulated relationship between level of arousal and performance. Since it has been suggested that time of day and sleepiness are related, we can suppose that in general there is a gradual rise in arousal from the morning into early evening. In Fig. 9 X and Y represent the levels of arousal early and later in the day respectively. These levels of arousal yield levels of performance E and L; thus performance will be better later in the day (it must be emphasized that one would not, under normal conditions, expect the arousal level to reach a* **post**-*optimal value at any time during the day when a task such as letter cancellation is being performed; however, it does not matter exactly where the "early" and "late" readings are placed on the arousal continuum, since the predictions we shall make are qualitative).*

*We now cause a general rise in arousal level by introducing an arousing factor like KR or noise, such that X moves to Xa and Y moves an equal distance to Ya. The resulting changes in performance are from E to Ea and from L to La; an improvement occurs in both cases, but the former greatly exceeds the latter. Thus, when an arousing factor is introduced, we can expect (i) that performance early in the day will be enhanced, but (ii) that performance later in the day will be enhanced to a far smaller extent, or even (if arousal is already at the optimum level) be slightly* **reduced.** *In short, the "time of day" effect will be eliminated by arousing factors, chiefly through the induction of high levels of performance early in the day. The following 2 experiments conducted by the author of this chapter in effect provide a test of this hypothesis.*

*Experiment 9: Letter Cancellation with* KR

Thirty subjects were used in this experiment. The task and procedure were exactly the same as described in the report of Experiment 1, with the sole exception that, at the end of each test session, the output scores of all the subjects in the current test group (normally 6) were first announced verbally and then posted prominently for each subject to inspect at leisure.

Analysis of variance showed that the time of day effect was not statistically significant for either the output score ($F = 0.74$) or the error score ($F = 1.12$). It will be remembered that in Experiment 1 there **was** a statistically significant diurnal variation in the output score; it would thus appear that the time of day effect on output was effectively abolished by the provision of $KR$. The actual mean output scores obtained in the present experiment are plotted in Fig. 10, in which, for comparison, the corresponding scores from Experiment 1 are also shown.

$KR$ appeared to increase output generally, but to different extents at the different testing times. The greatest increase was at 08.00 (10·1%), and the least at 21.00 (0·2%), i.e., there seemed to be an interaction between the $KR$ effect and time of day.

*The direction of this interaction was consistent with the hypothesis under examination. Since there was nothing to suggest that the 2 groups of subjects were not comparable (they were both drawn from the same parent population, and there was no significant difference in their introversion scores, or in the extent to which their temperature rose between 08.00 and 21.00), a test of the statistical significance of this interaction was made in a combined analysis of variance of the scores from both experiments. The interaction was found to be significant in this analysis at a high level of confidence ($F = 3.61$, $p < 0.01$) (the corresponding interaction in the case of the error scores was not statistically significant ($F = 0.61$).*

*Experiment 10: Letter Cancellation in noise*

Although the task was exactly the same, this experiment differed in design from the one employed in the $KR$ study. Subjects were tested at 2 times of day (08.00 and 10.30) but under both "noise" and "quiet" conditions; each subject performed once under each combination of factors. In the "noise" condition continuous thermal noise was relayed over a loudspeaker at a level of 100 dB throughout the test session. In

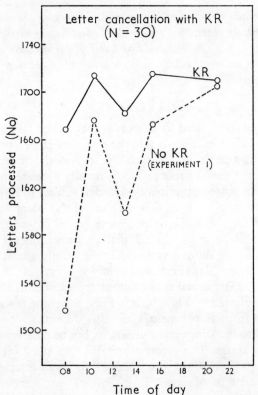

Fig. 10. Mean output scores in the Letter Cancellation with *KR* test (Experiment 9). The corresponding scores from Experiment 1 are also shown.

the "quiet" condition the noise was switched off, but ventilating fans provided a background level of approximately 70 dB. Twenty-one subjects and 4 different orders of testing were employed. Fifteen subjects were assigned in equal number to 3 of these orders; 6 subjects were assigned to the remaining order. Overall mean scores under the 4 combinations of experimental variables are shown in Table XI.

Both the output and the error scores were higher in noise than in quiet, but the only difference found to be statistically significant was the 6·3% increase in output at 08.00 ($p < 0.05$, Wilcoxon test). A slight rise in mean temperature in the noise condition was also observed, but this was not statistically significant at either test time.

Another way of describing these results is to say that the 9·8% increase in output between 08.00 and 10.30 under quiet conditions was reduced in the presence of noise to 3·6%. Since the corresponding increases for the

TABLE XI

*Letter Cancellation in noise: mean performance scores over all subjects*

|          | 08.00 Output | 08.00 Error | 10.30 Output | 10.30 Error |
|----------|--------------|-------------|--------------|-------------|
| "Quiet"  | 1194         | 20%         | 1311         | 1·7%        |
| "Noise"  | 1269         | 23%         | 1315         | 2.4%        |

no-*KR* and *KR* conditions over the same period were very similar (10·6% and 2·6% respectively), it would appear that *KR* and noise were affecting performance in much the same way.

*These new results, taken in conjunction with those described earlier, support very strongly the notion that time of day effects are the effects of variations in* **sleepiness** *or* **arousal** *during the day. The extent to which the observed differences in performance in introverts and extraverts are explicable in these terms must now be examined.*

*The hypothesis stated in Corcoran's studies that introverts are characterized by higher levels of arousal than extraverts must be taken to imply that introverts* **always** *have the higher level,* **not** *that their relative levels change as the day progresses. Yet we have seen that the main difference in the circadian temperature rhythms of groups of introverts and extraverts is one of* **phasing***, which results in a crossover of the curves during the day. Thus if it is held that temperature is a valid indicant of level of arousal, then we must state the alternative hypothesis that* **introverts have higher levels of arousal in the morning, but that at later times of day, extraverts have the higher level.**

*These 2 hypotheses can be tested upon the data from the 3 Letter Cancellation experiments. Both hypotheses predict that the addition of* KR *or noise will have a greater effect upon extraverts than upon introverts at 08.00, when the initial level of arousal is relatively low. At later testing times, when the initial level is closer to the optimal, Corcoran's theory predicts that the addition of these arousing factors will have if anything a* **deleterious** *effect upon introverts, but little or no effect upon extraverts; the alternative hypothesis on the other hand suggests that* KR *and noise will affect* **extraverts** *deleteriously rather than introverts.*

*Figure 11 shows the adjusted ouput scores of extraverts (N = 14) and introverts (N = 10) in Experiment 9 (KR); also plotted are the scores from the corresponding groups in Experiment 1 (no KR). At 08.00 KR affected extraverts beneficially (t = 4·71, p < 0·01), as both hypotheses predict, but at 21.00* **extraverts were affected deleteriously by** KR (t = 3·18, p < 0·01). *By contrast, introverts were not significantly affected by KR at* **any** *time. These results, which are reflected in the direction of the correlations between introversion rating and adjusted output at the different testing times (see Table XII) yield clear support for the alternative hypothesis.*

TABLE XII

*Letter Cancellation with KR: correlations between adjusted performance, adjusted temperature and introversion at each time of testing*

| | Introversion × Temperature | | | | |
|---|---|---|---|---|---|
| | 08.00 | 10.30 | 13.00 | 15.30 | 21.00 |
| | +0·307* | +0·245 | −0·093 | −0·053 | −0·354* |

| | Introversion × Performance | | | | |
|---|---|---|---|---|---|
| | 08.00 | 10.30 | 13.00 | 15.30 | 21.00 |
| Output | −0·209 | +0·131 | −0·189 | +0·134 | +0·362* |
| Error | −0·112 | −0·211 | +0·101 | +0·316* | +0·121 |

| | Temperature × Performance | | | | |
|---|---|---|---|---|---|
| | 08.00 | 10.30 | 13.00 | 15.30 | 21.00 |
| Output | +0·284 | +0·089 | +0·110 | −0·103 | −0·095 |
| Error | −0·008 | −0·452† | −0·124 | +0·048 | −0·075 |

*$p$ (one-tailed) $< 0·05$
† $p$ (one-tailed) $< 0·01$

The adjusted temperature curves for introverts and extraverts in the KR experiment are shown in Fig. 12. From these curves, and the patterning of the correlations between introversion and temperature through the day (Table XII) it is clear that the present group of subjects was comparable both to the original large sample and to the group tested without KR in Experiment 1. In each case introverts had relatively higher temperatures at 08.00, and at 21.00 the reverse was true.

Table XIII shows the mean output scores of introverts and extraverts in Experiment 10 (noise) (note: because of the relatively small number of subjects tested in this experiment, individuals with a Heron Inventory score of 4 were classified as "introverts"; thus all subjects were included in this analysis). In noise there was a statistically significant **increase** in output at 08.00 in the extravert group ($p < 0·01$, Wilcoxon test) but not in the introvert group, whereas at 10.30 the output of the extraverts actually **declined**. This decline was not, however, statistically significant. It is perhaps unfortunate that the "later" time chosen in this experiment was 10.30 rather than 21.00, since clearer results might have been obtained in the latter case; nevertheless, we can say that the present findings are at least **generally** supportive of the alternative hypothesis.

Taken altogether, the results described in this chapter therefore favour the view
(1) that introverts have higher arousal levels than extraverts in the morning,
(2) that there is a general increase in level of arousal in both "types" throughout the day,

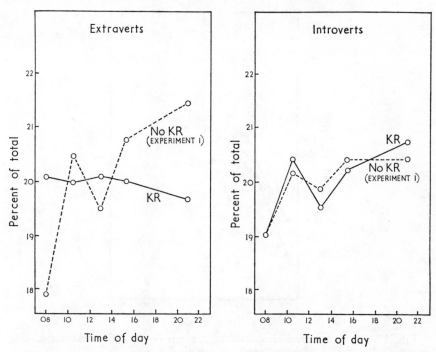

FIG. 11. Mean adjusted output scores for introverts and extraverts in the Letter Cancellation with *KR* test. The corresponding scores from Experiment 1 are also shown

(3) *that the level of arousal increases at a greater rate in extraverts than in introverts, with the result that*

(4) *when an additional arousing factor is added to the task situation the level of arousal in extraverts may be post-optimal for performance later in the day.*

Changes in temperature during the day mirror the performances of introverts and extraverts quite neatly under "normal" conditions, and could therefore (apart from the divergent trends in the post-lunch period) be considered a reasonably valid indicant of variations in arousal, and in resultant levels of efficiency, in both types.

TABLE XIII

Letter Cancellation in noise: mean output scores for introverts (N=10) and extraverts (N=11)

|  | 08.00 | | 10.30 | |
|---|---|---|---|---|
|  | Introverts | Extraverts | Introverts | Extraverts |
| "Quiet" | 1237 | 1149 | 1364 | 1253 |
| "Noise" | 1256 | 1283 | 1405 | 1216 |

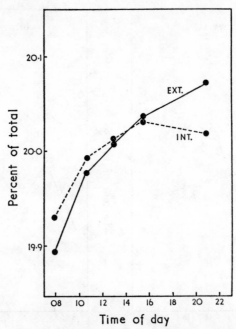

FIG. 12. Mean adjusted temperatures for introverts and extraverts in the Letter Cancellation with *KR* test.

*However, because of the "inverted-U" relationship between performance and arousal it can now be seen that the use of temperature as a simple predictor of actual behaviour will, in certain circumstances, be misleading. It was noted in Experiment 1 (Letter Cancellation under "normal" conditions) that the sign of the correlation between temperature and output remained positive throughout the day (see Table II). However, when the same task was performed under "arousing" conditions (Experiment 9, KR) this consistent positive correlation was replaced by a trend from positive to* **negative** *values as the day progressed (see Table XII). It seems reasonable to suppose that these negative values resulted from the fact that KR induced a state of* **hyper-arousal** *in the extraverted subjects later in the day which affected their performance adversely (cf. Fig. 11). Clearly, then, the value of temperature as a direct index of performance efficiency is limited to conditions in which arousal does not exceed the optimal level for the particular task under investigation.*

*Finally, it is of interest to consider the somewhat unexpected results of the Digit Span experiment in the light of the hypothesis that arousal increases during the day. In general there are 2 theories about the loss of material in short-term memory (see Melton, 1963). One states that memories can decay autonomously with time, the other that the loss of stored items is due to interference from subsequent and prior material. It seems unlikely that increased arousal could lead to faster passive decay of stored information. But it is conceivable that the increase in neural facilitation which is postulated to occur with heightened arousal (e.g., Bremer, 1961) might increase the effective strength of extraneous stimulation, and thus result in greater memory loss due to interference. Since performance in this short-term memory task deteriorated with increased levels of arousal, it would seem that the results therefore tend to favour the interference theory.*

## References

Bell, C. R. and Watts, A. N. (1966). *Brit. J. Psychol.* **57**, 155–159.

Blake, M. J. F. (1967). *Nature (London)* **215**, 896–897.

Bremer, F. (1961). *In* "The Nature of Sleep" (G. E. W. Wolstenholme and M. O'Connor, eds), pp. 30–50. J. and A. Churchill, London.

Colquhoun, W. P. (1960). *Ergonomics* **3**, 377–378.

Colquhoun, W. P. and Corcoran, D. W. J. (1964). *Brit. J. Soc. Clin. Psychol.* **3**, 226–231.

Corcoran, D. W. J. (1962a). "Individual Differences in Performance after Loss of Sleep". Unpublished Ph.D. thesis, University of Cambridge, England.

Corcoran, D. W. J. (1962b). *Quart. J. Exp. Psychol.* **14**, 178–182.

Corcoran, D. W. J. (1964a). *Brit. J. Psychol.* **56**, 267–273.

Corcoran, D. W. J. (1964b). *Amer. J. Psychol.* **77**, 298–300.

Davies, D. R. and Hockey, G. R. J. (1966). *Brit. J. Psychol.* **57**, 381–389.

Fisher, R. A. and Yates, F. (1953). "Statistical Tables for Biological, Agricultural, and Medical Research". Oliver and Boyd, Edinburgh.

French, J. W. (1942). *J. Exp. Psychol.* **31**, 79–87.

Heron, A. (1956). *Brit. J. Psychol.* **47**, 243–251.

Hoagland, H. (1933). *J. Gen. Psychol.* **9**, 267–287.

Kleinsmith, L. J. and Kaplan, S. (1963). *J. Exp. Psychol.* **65**, 190–193.

Kleitman, N., Titelbaum, S. and Feivson, P. (1938). *Amer. J. Physiol.* **121**, 495–501.

Kleitman, N. (1963). "Sleep and Wakefulness". University of Chicago Press, Chicago.

Leonard, J. A. (1959). *Med. Res. Counc. Appl. Psychol. Res. Unit.* Report No. 326., Cambridge, England.

Mackworth, N. H. (1948). *Quart. J. Exp. Psychol.* **1**, 6–21.

Melton, A. W. (1963). *J. Verb. Learn. Verb. Behav.* **1**, 1–21.

Pfaff, D. (1968). *J. Exp. Psychol.* **76**, 419–422.

Poulton, E. C. (1966). *Annu. Rev. Psychol.* **17**, 177–200.

Poulton, E. C. and Freeman, P. R. (1966). *Psychol. Bull.* **66**, 1–8.

Siegel, S. (1956). "Nonparametric Statistics for the Behavioural Sciences". McGraw-Hill, New York.

Terman, L. M. and Merrill, M. A. (1937). "Measuring Intelligence". Harrup, London.

Thor, D. H. (1962). *Percept. Mot. Skills.* **15**, 451–454.

Wilkinson, R. T. (1958). "The Effects of Lack of Sleep on Perception and Skill". Unpublished Ph.D. thesis, University of Cambridge, England.

Wilkinson, R. T. (1961). *J. Exp. Psychol.* **62**, 263–271.

Wilkinson, R. T. (1963). *J. Exp. Psychol.* **66**, 332–337.

Wilkinson, R. T. (1964). *Ergonomics*, **7**, 63–72.

Wilkinson, R. T. (1965). *In* "The Physiology of Human Survival" (O. G. Edholm and A. L. Bacharach, eds), pp. 399–430. Academic Press, New York and London.

Williams, H. L., Lubin, A. and Goodnow, E. L. (1959). *Psychol. Monogr.* **73**, 1–26.

CHAPTER 4

# Sleep Behaviour as a Biorhythm

WILSE B. WEBB

*Department of Psychology, University of Florida, Gainesville, Florida, U.S.A.*

## I. Introduction

Whenever a psychologist's aim exceeds his experimental grasp of a particular problem he almost inevitably runs into a veritable bramble patch of conceptual problems. Examples past and present abound from philosophical issues of the Humpty Dumpty mind–body problem to the current social issues surrounding behaviour control. Sleep research seems particularly bedevilled by conceptual difficulties. The intimate and profound relationships of sleep to consciousness, the complexities of its biochemical and physiological interactions, and its overwhelming effects relative to environmental sensitivity and response capacity pose complicated problems of meaning. Even the specified context of biological rhythms presents fundamental issues. Is sleep a penumbra of a more fundamental rhythm or is it a major determinant of the rhythm itself? Does the organism wake from sleep to survive, or sleep to survive its waking? Are these discreet, interactive entities or are they expressions of a single continuum?

I shall conclude this chapter with some groping attempts to speak of the role of sleep in the biorhythm tide. I shall begin, however, on the safer grounds of data concerning the appearance of sleep across the circadian span. My conceptual assumptions will be quite meagre, and my approach generally operational. Specifically, I am defining sleep by changes in the electroencephalogram (EEG), thus evading problems of awareness and consciousness. I shall view the presence of sleep simply as a response on the part of the organism in much the same way that the eating of food or the drinking of water may be viewed as a response, without assumptions as to the state variables such as hunger or thirst or the substrata of these responses such as biochemical or physiological events.

It is unlikely that every reader of this volume is familiar with the particular problem of sleep and its measurement, so a brief review of the EEG as an index of sleep is in order. This is the most widely accepted method of measuring sleep and has 3 critical advantages. This measure evades the problem of awareness or consciousness; it may be utilized across species; and, most critically, it presents sleep as a continuous process. The limitations of other methods of measurement such as thresholds and motility have been reviewed elsewhere (Webb and Agnew, 1969).

The EEG was essentially conceived by the German psychiatrist Berger in 1929, and developed in the early 1930s as an experimental tool by Adrian and Matthews of Cambridge. It was firmly introduced into sleep research by Loomis, Harvey and Hobart when they noted that not only profound changes occurred in the EEG with the onset of sleep but clearly discernible and often changing characteristics marked the recordings throughout sleep (1935). A systematic coding of these changes was introduced by these authors labelled A through E (1937). After the stage of dreaming was recognized by Aserinsky and Kleitman (1953), a revised code was presented by Dement and Kleitman (1957). This basic coding has been articulated in detail by a committee report of the Association for the Psychophysiological study of Sleep (Rechtschaffen and Kales, 1968).

In brief, the presently accepted coding of human sleep includes a waking stage (W), 4 stages of non-dream sleep (stages 1 through 4), and dream sleep (REM). The characteristics of the tracings of these records are shown in Fig. 1. Descriptively, the stages are characterized as follows.

Stage W (waking); alpha activity (8–12 Hz waves) and/or low voltage mixed frequency EEG. Although some persons show no alpha activity in their waking records experimental subjects are often chosen for alpha presence to aid in scoring the W state.

Stage 1: relatively low voltage, mixed frequency EEG with a prominence of activity in the 2–7 Hz range.

Stage 2: the presence of "sleep spindles" and/or K complexes. Sleep spindles are regular 12–14 Hz "bursts" of waves and K complexes are sharp negative waves which are immediately followed by a positive component.

Stage 3: at least 20% and not more than 50% of the scoring period containing relatively high voltage waves of 2 Hz or slower.

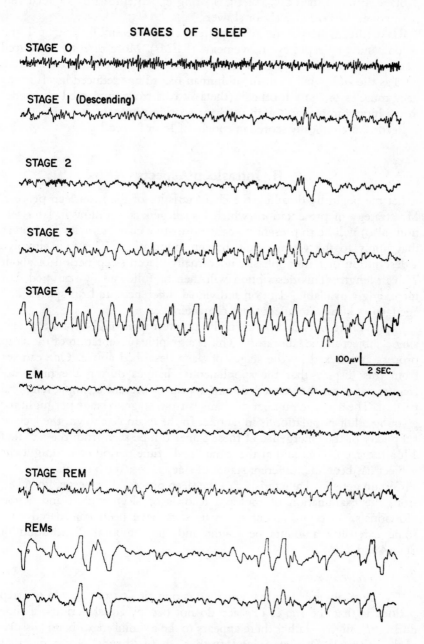

FIG. 1. Tracings of EEG stages of sleep.

F

Stage 4: more than 50% of the scoring epoch consisting of relatively high voltage waves of 2 Hz or slower.

REM (dream sleep): the EEG pattern closely resembling Stage 1 and accompanied by rapid eye movements (REMs). More recently a reduced electromyogram (EMG) tracing in the mental-submental area is often used as the additional sign. In subhuman recordings reduced levels in the neck muscles or "saw toothed" (theta) waves from implanted electrodes in the hippocampus are often used for scoring purposes.

Records are typically scored in epochs of 15 or 30 sec.

## II. Intrasleep Aspects

Let me begin by detailing the characteristics of the intrasleep process. My strategy of presentation (which I shall generally follow in later sections also) will be to present first the normative characteristics of sleep as they apply to the young adult human population. This will involve a description of a monophasic, nocturnally placed sleep period of about $7\frac{1}{2}$ h in length. This description will then be followed by scattered, but increasingly available, data on individual differences and on species and age factors known to affect these characteristics.

Figure 2 presents 3 experimentally uninterrupted nights of sleep of a single subject in the laboratory. This is a graphic presentation of the sleep process as indexed by the stages of sleep described above. One can see from this display that the spontaneous changes during the night are frequent and, on the surface, do not appear to present a simple temporal pattern. The average number of changes from stage to stage per night for a young adult population is in fact 32. In the example shown, the lack of simplicity of the emergence of these stages can be seen from the fact that 1 h after sleep onset and at the same "real" time period of midnight the subject has been in 3 different stages of sleep across the 3 nights.

Certain general characteristics of the sleep process do emerge, however, from the examination of multiple records of sleep within and across individuals. Two prominent characteristics have been considered with some extensity: amounts per night and the temporal distribution of State 4 and REM.

### A. *Sleep stage amounts*

The most obvious fact that emerges as one views a population of sleep periods is that the stages of sleep do not occupy equal periods of time during the night—rather there appears to be a "mini-maxi band" associated with each sleep stage and further, these "bands" have different

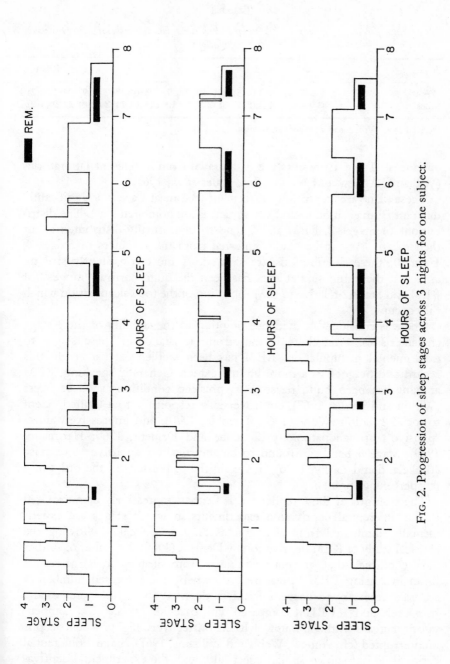

FIG. 2. Progression of sleep stages across 3 nights for one subject.

TABLE I

*Sleep stage percentage means and standard deviations (in parentheses) for two young adult populations*

| Stage | | 0 | 1 | 2 | 3 | 4 | REM |
|---|---|---|---|---|---|---|---|
| WAW | $N = 16$ | 1·0 (0·9) | 5·4 (4·5) | 48·7 (4·7) | 7·7 (3·0) | 13·2 (5·9) | 24·1 (4·2) |
| Kales | $N = 16$ | — | 4·4 (2·9) | 51·2 (5·8) | 10·1 (3·6) | 11·2 (4·8) | 22·9 (3·9) |

Derived from Williams *et al.* (1964) and Kales *et al.* (1967).

variances. Table I presents the data drawn from 2 different laboratories (Williams, Agnew and Webb, 1964; Kales *et al.*, 1967).

These data are in striking agreement (the means are not significantly different from each other and the variances are homogeneous). The distributions of Stages 2, REM and 3 tend to be normally distributed about their means. As can be readily inferred from the variances of Stages 4, 1 and 0, these distributions are skewed. A most consistently striking aspect of these and similar data has been the quite narrow "ranges" of REM and Stage 2 relative to the proportion of the total sleep period which they occupy.

There is considerable evidence pointing to the stability of these stage characteristics. First there is the strong resistance of these stages to experimental manipulation. REM has been studied extensively in this regard and Stage 4 to a more limited extent. Generally speaking, REM amount has been quite resistive to presleep conditions such as stress (Witkin and Lewis, 1967), controversially responsive to a limited extent to social isolation (Wood, 1962; Tremble, 1970), and impressively unresponsive to psychopathology (Zarcone and Dement, 1969; Hartmann, 1969). Stage 4 has been found to be statistically responsive to exercise (Baekeland and Lasky, 1966), but again the response is in the order of a few percentage points.

The strength of these tendencies is further impressively demonstrated in the differential deprivation experiments in which stages are experimentally eliminated from sleep for as long as a week. Following the classical study of REM deprivation by Dement (1960), a number of studies have confirmed that preventing subjects from sleeping in REM results in an increasing REM "pressure", shown by the increased number of arousals necessary to prevent REM as deprivation nights increase, and by a "rebound" of REM compared with baseline levels, i.e., a compensatory increase in the amount of REM on recovery nights when sleep is uninterrupted (cf., Agnew, Webb and Williams, 1967). Stage 4 differential deprivation has shown similar effects although the experimental depriva-

TABLE II

*Average correlations of sleep stage percentages for nights 2–3 and 3–4 of laboratory sleep conditions by age groups*

| Age group | 0 | 1 | Sleep stage REM | 2 | 3 | 4 |
|---|---|---|---|---|---|---|
| 8–11 yr | 0·07 | 0·42 | 0·42 | 0·69 | 0·75 | 0·50 |
| 16–19 | 0·19 | 0·31 | 0·26 | 0·60 | 0·14 | 0·48 |
| 20–29 (m) | 0·02 | 0·62 | 0·48 | 0·52 | 0·60 | 0·70 |
| 20–29 (f) | 0·07 | 0·31 | 0·54 | 0·72 | 0·73 | 0·75 |
| 30–39 | 0·44 | 0·42 | 0·35 | 0·75 | 0·26 | 0·76 |
| 50–59 | 0·39 | 0·60 | 0·58 | 0·84 | 0·70 | 0·70 |

tion "pressure" seems higher and the recovery effects more attenuated (Agnew *et al.*, 1967).

The presence of stable individual differences from night to night lends evidence to the stability of sleep stage amounts over a full night. The variances given in Table I clearly indicate a range associated with each stage. The evidence that these are not merely reflecting an error variance may be seen in Table II which presents the between-night correlations for a broad age range of subjects. It is obvious from this table that the differences between subjects is a consistent phenomenon. The low correlations associated with Stages 0 and 1 in the younger subjects undoubtedly reflect the transient nature of this variable in these age ranges. This table presents evidence that a stable estimate of an individual's "characteristic" stage amounts could be derived from as few as 3 nights of recordings for most stages and at the most, 6 nights of recordings for all stages.

The process of ageing has a profound effect on the total sleep stage amounts which occur during a night. The extent and character of this effect may be seen in Table III.

These data from our laboratory parallel those reported by Feinberg and Carlson (1968) on 5 age groups ranging from 6 to 84 years of age. To this table should be added the consistent estimate of REM in the neonate as approximating 50% of the sleep period (Roffwarg, Muzio and Dement, 1966), and the further reduction of Stage 4 and the probable slight reduction of REM in more elderly subjects (Kales *et al.*, 1967; Kahn and Fisher, 1969). In general, we may say that age has a strong effect on REM between the ages of birth to about 10 years of age and a possible limited effect beyond the age of 60. On the other hand, Stage 4 shows a marked reduction as a function of age beginning in the mid-thirties, with an associated increase in Stage 0 and Stage 1.

In reference to comparative data on sleep stage amounts, the picture is somewhat complicated by differences in EEG characteristics in lower animals. The amount of REM sleep has been extensively studied, however,

TABLE III

*Sleep stage percentages for age groupings*

| Subject group | 0 | 1 | REM | 2 | 3 | 4 | Mean sleep length (min) | No. Ss |
|---|---|---|---|---|---|---|---|---|
| 21–31 mo. | 2 | 8 | 29 | 43 | — | 18 | 596 | 16 |
| 8–11 yr | 2 | 6 | 24 | 44 | 6 | 18 | 565 | 18 |
| 16–19 | 1 | 5 | 23 | 47 | 6 | 18 | 454 | 32 |
| 20–29 (m) | 1 | 5 | 24 | 50 | 7 | 13 | 418 | 28 |
| 20–29 (f) | 1 | 6 | 22 | 48 | 7 | 16 | 451 | 16 |
| 30–39 | 2 | 8 | 22 | 53 | 5 | 10 | 443 | 12 |
| 50–59 | 4 | 11 | 23 | 51 | 8 | 3 | 436 | 16 |
| 60–69 (m–f) | 9 | 12 | 20 | 51 | 5 | 3 | 440 | 16 |

and should be noted. In general, it has been found that the EEG pattern of sleep in the primate is similar, if not identical, to that of human sleep (cf., Weitzman *et al.*, 1965; Kripke *et al.*, 1968). Below the primate, however, the scoring of sleep has generally been in the categories of non-REM and REM sleep. Ursin (1968), however, has described a 2-stage non-REM sleep in a cat, and Van Twyver (1967) noted a "somnolent" sleep in the grayline ground squirrel. Non-REM sleep (being characterized by relatively high amplitude waves) and REM sleep (characterized by cortical EEG activity resembling the awake state, sub-cortical synchronized waves from the hypothalamic region, and associated physiological responses such as lowering of muscular tonus) appear to be universal in the mammalian order. In birds (Klein, Michel and Jouvet, 1964) non-REM sleep is present but REM sleep is quite minimal (less than 1% of the sleep period). In the turtle, although slow sleep may be detected, there is no REM sleep (Hermann, Jouvet and Klein, 1964). Sleep can be detected in the lizard, and there appears to be an equivalence of REM sleep, but the EEG characteristics are different from those found in mammals (Tauber, Roffwarg and Weitzman, 1966). In a recent study, no EEG sleep or REM sleep could be detected in 1 type of frog (Hobson, 1967).

The percentage of the total sleep period that REM time occupies varies quite widely across species. As noted above, it does not appear in the turtle; it is less than 1% in the bird. In mammals, the percentage of REM sleep relative to total sleep in the animals so far measured varies from about 30% in the opossum (Synder, 1964) to about $2\frac{1}{2}$% in the sheep (Ruckebusch and Bost, 1962). The mole, ground squirrel, cat and man all spend between 20–25% of their sleep time in REM states. Within a phylum the proportion of REM sleep may vary widely. Among the rodents, REM sleep will vary from 5% in the guinea-pig (Pellet and Beraud, 1967) to 15% in the laboratory rat to 25% in the hamster (Van Twyver, 1967).

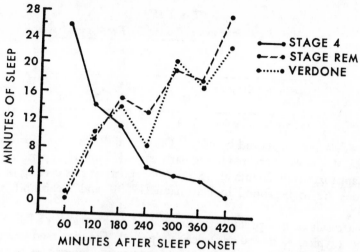

FIG. 3 Hourly distribution of Stage 4 and Stage REM.

B. *Temporal aspects*

In spite of an apparent chaotic emergence of stages across the night displayed in Fig. 2, 2 systematic temporal characteristics can be discerned in a population of sleep measures: a basic sequential relationship and a differential temporal distribution of Stage REM and Stage 4.

Stages do not emerge in a random order, but are in a sequential ordering. In general, there is an ordinal relationship from Stages 1 through 4 progressing from 1 to 2 to 3 to 4, and in reverse from 4 to 3 to 2 to 1 (Williams *et al.*, 1964). Exceptions to this order may occur in the 4 to 1 sequence due to arousals resulting from internal or external stimulation. This sequence is also independent of REM which seems to be linked with Stage 2 since it is preceded or followed by Stage 2 in about 85% of its occurrences.

A second temporal patterning of the sleep process is seen in the differential distribution of the amount of Stages 4 and REM which occurs across the night. Figure 3 presents hour by hour amounts of Stages 4 and REM recorded from 32 young males (3 nights each) from our laboratory (Webb and Agnew, 1969), as well as REM amounts from Verdone (1968). Clearly, Stage 4 appears predominantly in the early part of the sleep period and REM in the latter part. This differential distribution has also been noted by Kales in a first half–last half of sleep period analysis (Kales *et al.*, 1967).

Although Fig. 3 gives the appearance of a graduated decrease of Stage 4 and an increase of REM per hour, an important underlying characteristic

158                                    WILSE B. WEBB

TABLE IV

*Mean length in minutes of successive REM periods*

|                      | Period | | | | |
| --- | --- | --- | --- | --- | --- |
|                      | 1  | 2  | 3  | 4  | 5  |
| Dement and Kleitman  | 9  | 19 | 24 | 28 | 34 |
| Verdone              | 13 | 27 | 30 | 33 | 31 |

Derived from Verdone (1968).

of this distribution should be noted. Dement and Kleitman (1957) noted
that REM episodes occurred in regularly spaced "bursts" or periods which
were approximately 90 min apart. Further reports of this cycling of REM
episodes are to be found in Hartmann (1968)* and Webb and Agnew
(1969).

As a result of this periodicity, there are 4 to 5 periods of REM sleep
during a typical 7½ h (450 min) sleeping period. The increased amount of
REM across the night then results from increasing lengths of successive
REM episodes. Data presented in Table IV taken from the Verdone
study (1968), which also includes the data from the earlier study by Dement
and Kleitman (1957), document this characteristic of REM sleep.

Stage 4 shows a similar, although not equally systematic, episodic
character. Typically, Stage 4 will have a latency of approximately 30 to
45 min from sleep onset, and will last between 15 and 30 min. This episode
will often be interrupted briefly by Stage 3 or Stage 2. A second episode
will follow between about 60 and 90 min and typically last for a period
shorter than the first episode. If there is a third episode, it is likely to be
quite transient. There are seldom more than three such periods (Webb
and Agnew, 1969).

As with total sleep stages, we have been impressed by the stability or
"toughness" of this intrasleep characteristic. We first noted the resistance of this pattern to experimental manipulation in an earlier study
of restricted sleep regimes (Webb and Agnew, 1965). In this study,
subjects were permitted to sleep only 3 h per night for 7 days. Table V
compares the sleep which occurs in the first 3 h during a full night's
sleep (baseline nights) and sleep occurring in the restricted period.

It will be noted that in spite of severe REM deprivation, the restricted
sleep period continued to be dominated by Stage 4 with limited REM
sleep as would be predicted by Fig. 3. Three subsequent studies of restricted sleep (Sampson, 1966; Clemes and Dement, 1967; Greenberg et al.,
1968) generally confirm the results presented in Table V. A tendency for

* Hartmann calculated the sleep cycle from the end of one REM period to the end of the
next. We prefer to define this period from the onset of a REM cycle to the onset of the next
REM cycle.

TABLE V

*Percentage sleep stage amounts for first 3 h of baseline night sleep and 7 nights of 3 h per night sleep*

|  | Stages of sleep | | | | | |
| --- | --- | --- | --- | --- | --- | --- |
|  | 0 | 1 | 2 | 3 | 4 | REM |
| First 3 h baseline | 1 | 4 | 37 | 12 | 38 | 8 |
| Restricted sleep | 1 | 3 | 35 | 7 | 48 | 8 |

Stage REM to move forward in time, i.e., to occur more predominantly in the early part of sleep, was more notable, however in these latter studies. Even more impressive is the recent report of Jones and Oswald (1968) of 2 cases who had slept about 3 h per night for 20 years and 6 years, respectively. It was found that about 50% of their sleep was in Stages 3 and 4 and there occurred about 40 min of REM in the 165-min sleep periods recorded. This amount would be predicted if chronic REM deprivation had moved this tendency forward in time sufficiently to permit 2 "bursts" of REM sleep.

Two additional studies speak directly to the stability of this differential distribution of Stages 4 and REM. Figure 4 presents a proportional distribution of arousals required to eliminate Stage 4 or Stage REM relative to thirds of experimental nights in which Stage 4 or REM was being eliminated (Williams, Agnew and Webb, 1967). It is clear from this figure that more intense arousal efforts are required for Stage 4 early in the night and for REM later in the night.

The other study attempted to displace Stages 4 and REM into their less predominant temporal periods (Agnew and Webb, 1968). Results of this study may be quoted directly: "When an attempt was made to force Stage 4 sleep to occur during the last third of sleep and REM to occur during the first third, it was found that the procedures were only partially successful. When Stage 4 was permitted to occur only during the last part of the sleep period, an average of 44 min of Stage 4 occurred as compared with a baseline average value of 87 min for a full night. Similarly, an average of 21 min of REM sleep occurred during the first 2 h of sleep as compared with a baseline average of 71 min for a full night" (p. 147).

Figures 5A and 5B present the curves of Stages 4 and REM as a function of time asleep for 3 age groups. It is clear from these figures that the distribution holds across age ranges, with only the slope of the curve varying as a result of different amounts of the stages as a function of age.

As noted above relative to stage amounts, in animals lower than the primate, sleep records are scorable into only 2 categories and hence

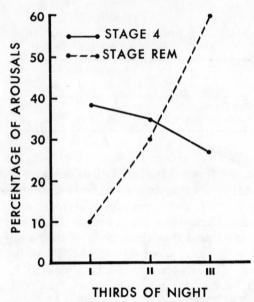

FIG. 4. Percentage of arousals required to eliminate Stages 4 and REM during thirds of night.

sequential data and relative temporal distributions within sleep periods are not parallel to the more complex human data. However, the length of the interval between REM cycles has been studied in some detail. The average cycle lengths in minutes quoted by Hartmann (1967) are as follows: mouse (3–4), rat (7–13), rabbit (24), opossum (17), cat (20–40), monkey (40–60), man (80–90) and elephant (120). Kleitman, in a review of the basic rest-activity cycle (1967), attributes this increase in duration to body size. Hartmann (1967) has attributed the change to metabolic rate.

A recent study of the cycling of sleep in the rat (Webb and Friedmann, 1969) indicates one striking difference which may characterize lower animals relative to the sleep of humans in temporal distribution aspects. This study noted that the probability of a REM episode was related to the amount of preceding sleep. It will be recalled that the rat only exhibits slow wave sleep and REM sleep so that this finding can be interpreted as roughly paralleling the differential distribution of slow wave sleep preceding REM sleep as seen in the human subject. It was found, however, that although the time interval between successive REM episodes was essentially constant, there was not an increasing period length as a function of successive episodes as is seen in human subjects.

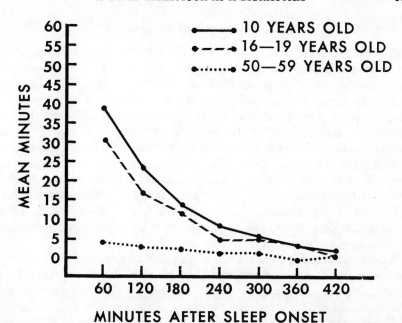

FIG. 5A. Amount of Stage 4 as a function of time asleep for 3 age groups.

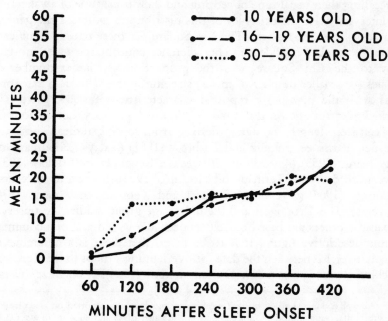

FIG. 5B. Amount of Stage REM as a function of time asleep for 3 age groups.

# III. Length of Sleep

Perhaps the most obvious aspect of sleep is the amount or length of time that it occupies during the 24 h. As such, it is quite surprising that this characteristic of sleep has received such limited attention. Kleitman (1963), for example, devotes 7 pages of his 370-page text to the topic of sleep duration and cites some 40 descriptive studies of "natural sleep length" in among the 4337 studies listed in his bibliography. Of these studies, nearly half were reports prior to 1930 and only 7 were subsequent to 1950. The topic fares even more poorly in the more recent books on sleep. In 12 current editions (Abt and Riess, 1969; Foulkes, 1966; Koella, 1967; Murray, 1965; Oswald, 1962; Hartmann, 1967; Wolstenholme and O'Connor, 1961; Jouvet, 1965; Luce and Segal, 1966; Webb, 1968; Kales, 1969; Kety, Evarts and Williams, 1967), there are only 2 references to sleep length or duration in the indexes.

Without attempting to account for this puzzling neglect, let me attempt to briefly summarize the available data on sleep length which I have briefly reviewed elsewhere (Webb, 1968, 1969).* Table VI presents the self-reports of 4375 17- and 18-year-old students entering the University of Florida concerning their average length of sleep in the previous year (Webb and Friel, 1970). They were asked to estimate this sleep duration by recalling their usual time of retiring and their usual time of awakening. An interview with the extremes of this distribution indicated the general accuracy of the self-reports. The Chi Square for these data is significant at the 1% level of confidence. The difference contributing most substantially to the Chi Square was the proportionately larger number of females (or smaller number of males) sleeping in the $5\frac{1}{2}$ to $6\frac{1}{2}$ h category.

As with the previously reported characteristics of sleep, there is a marked effect of age on this variable. Figure 6 presents the relationship between sleep length and age as derived from several studies. These are, in order, Parmalee, Schultz and Disbrow (1961) (3 days); Kleitman and Engelmann (1953) (4 weeks to 26 weeks); Reynolds and Mallay (1933) (2 years to 5 years); Terman and Hocking (1913) (6 years to 18 years); O'Connor (1964) (6 years to 65 years); and Webb and Swinburne (1970) (75 years). The 3 youngest and the oldest age group studies are observational; the others are based on self-reports. We would add one comment to this descriptive figure. It is to be noted that there is a difference of slightly over 1 h between the data for overlapping groups in the Terman–Hocking study (1913) and the O'Connor study (1964). We believe this is a

---

* A recent well conducted study using 2 week sleep logs has been reported on sleep lengths on a substantial population (509 subjects) between the ages of 20 and 80 years of age (Tune, 1969).

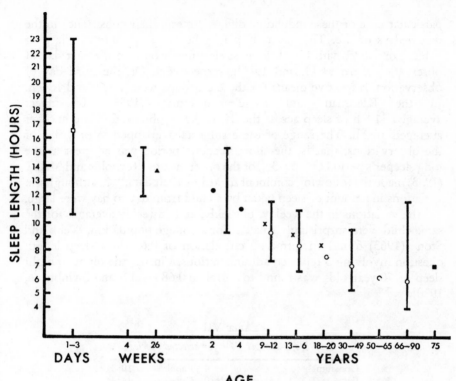

FIG. 6. Relationship between sleep length and age.

TABLE VI

*Entering 18 year old college students' self-reports of average sleep length*
*(percentage male and female)*

| | | 5½ h or less | 5½-6½ | 6½-7½ | 7½-8½ | 8½ or more |
|---|---|---|---|---|---|---|
| Male | (N = 2364) | 1·1 | 6·1 | 27·9 | 48·8 | 16·0 |
| Female | (N = 2011) | 0·6 | 9·2 | 31·1 | 44·5 | 14·5 |

real difference rather than an artifact. Both sets of data were derived from self-reports, but were obtained approximately 50 years apart. It is quite possible that children are sleeping 1 h less at the present time than they were sleeping 50 years ago.

The figure shows a wide range of individual differences throughout the various age groups. There is evidence to suggest that these are stable and systematic differences between subjects. Parmalee *et al.* (1961) made particular note of these individual differences and their consistency in the first 3 days of life. The mean sleep lengths of his 19 shortest sleeping babies compared with his 19 longest sleeping babies on the first day of observation were 14·4 h and 19·7 h, respectively. On the third day of observation, respective means for these 2 groups were 15·7 h and 16·9 h. In the Kleitman and Engelmann study (1953), 1 subject averaged 11·5 h of sleep across the 24 weeks of observation and another averaged 16·2 h. The range of these subjects overlapped in only 3% of the observations; that is, the short sleeper's period was as great as the long sleeper's period on only 3% of the observations. Reynolds and Mallay (1933) made the following comment regarding their data:". . . although the variations in amount of sleep taken by a child from day to day were large, . . . the variations in the weekly, biweekly, and triweekly averages for the same child were surprisingly small". For a college population, Webb and Stone (1963) found an intraclass correlation of 0·65 on a sleep length question involving 3 repeated administrations. Finally, the night to night sleep of 70-year olds was found to correlate 0·68 (Webb and Swinburne, 1970).

TABLE VII

*Length of sleep per 24 h for selected species*

| Opossum | 19·4 | Rabbit | 10·8 |
|---|---|---|---|
| Bat | 18·0 | Cat | 14·4 |
| Arctic Ground Squirrel | 16·5 | Tapir | 6·2 |
| Ground Squirrel (Grayline) | 13·8 | Sheep | 9·3 |
| Rat | 13·5 | Pig | 13·2 |
| Chimpanzee | 11·0 | Mole | 8·4 |
| Rhesus | 9·1 | Echnida | 12·0 |

Even more impressive than the onto-genetic influence on sleep length is the phylo-genetic effect. The most extensive analysis on the comparative data relative to sleep length has been completed recently by Zepelin and Rechtschaffen (1970), who have reviewed the data on 31 species. A sampling of sleep length across species from this study is given in Table VII.

## IV. Circadian Aspects of Sleep

To this point, we have been describing the intrasleep process as it occurs in the normal subject in the western cultural environment. For individuals beyond about 5 years of age (Webb, 1969) this is a uniphasic nocturnal pattern. Variations in the time placement of the sleep period

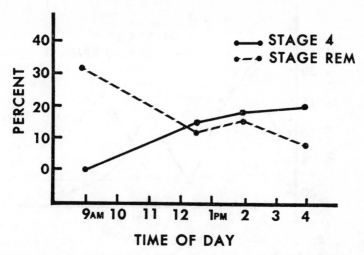

FIG. 7. Percentages of Stages 4 and REM present in 2-h sleep periods occurring during the day.

can occur as well as variations in number of periods. These variations will be reviewed in this section.

### A. *Temporal displacements*

We begin by considering the nature of sleep as it is "interjected" into different time periods within the basic sleep waking sequence. Simply, what is the nature of sleep behaviour when it begins at points in time other than its typical onset time which is approximately midway in the nocturnal period? Figure 7 presents the REM and Stage 4 amounts which are present in 2-h sleep periods the onsets of which occur after a full night's sleep at varying intervals.

To what may we attribute the systematic changes in sleep tendencies? We cautiously restated those results (Webb and Agnew, 1967) as follows: "If the interjected period approaches the onset of the regular period, the Stage 4 and 1 REM distribution will approximate the first portion of the regular sleep cycle. If, on the other hand, the interjected period is proximal to the termination of the regular period, there will be a dominant pattern of 1 REM with limited Stage 4 . . . (p. 159)." In fact, we believed that the increasing presence of Stage 4 was attributable to the time awake (and associated activity) and REM was a function of the amount of time (and need) development from the end of the previous sleep period. It is clear in retrospect, however, that some or all of the effect noted could have been simply an expression of a "time locked" circadian rhythm, which

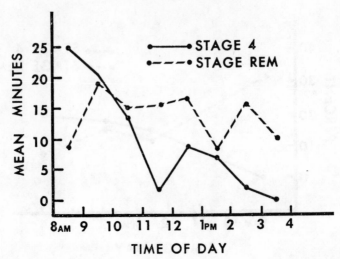

FIG. 8. Hourly amounts of Stages 4 and REM occurring during daytime sleep.

varied concomitantly with time awake or time distant from the termination of sleep.

Unfortunately, studies of the effect on sleep of its displacement into varying time periods within the 24 h are quite scarce. We know of only 2 quite recent experimental studies which have reported the characteristics of sleep when the regular cycle is displaced into other time periods (Weitzman et al., 1970; Agnew, Webb and Williams, 1968). Although sleep onset was at 10.00 in the Weitzman study and at 08.00. in our study, the findings agree in essential detail: the overall percentage of sleep in each stage was almost identical with baseline figures. There was, however, an increased number of arousals during the displaced sleep periods and, most critically, the temporal distribution of Stage REM and Stage 4 was quite distorted. The REM and Stage 4 amounts per hour across the displaced period are shown in Fig. 8. As can be seen in comparison with Fig. 5, the tendency to sleep in Stage REM is increased and Stage 4 tendencies are decreased early in the night whereas the reverse is true in the latter part of the sleep period.

In reviewing the other studies of phase shifts of the sleep period, we can find little reference to sleep itself. The Wyatt and Marriott study (1953) of factory shift work involving 555 workers reported that 42% of subjects on night shift were not satisfied with their sleep, and in the remaining 58% there were many qualifying remarks about daytime sleep compared with night-time sleep. Unfortunately, we have no control figures to indicate the number of people that were not satisfied with regular night sleep. In the remaining studies of time displaced sleep, we can find manifold reports of temperature, heart rate, performance, haematological data, urinary analysis, reaction time, and the like, but no further reference to sleep.

I know of no data on the effects of age on the sleep displacement variable. In the Agnew, Webb, and Williams experiment reported above, 2 different age groups were used, 20-year-olds and 30-year-olds. Unfortunately, the samples were too small to confirm differences which may have accrued from this small age difference. We are on record from indirect evidence as suggesting.that the modification of sleep patterns is likely to be a function of age, with the sleep patterns of the younger age groups being more readily modifiable (Webb and Agnew, 1962).

Although shifts in circadian rhythms have been extensively studied in lower animals, the primary variables observed have been activity and physiological rhythms such as temperature. We have been consistently unsuccessful in our attempts to train animals to displace their sleep response. We have used chronic periods of no sleep brought about by activity (Van Twyver and Webb, 1968; Friedmann, 1969) and shock contingently associated with the onset of the sleep response in set periods of time as well as noncontingent shock through set periods of time (Friedmann, 1969). We have found that as soon as the control system is removed, sleep returns to its normal rhythm. We have reported one study which suggests a strong circadian component in the sleep process in the rat. It was found that, when administered during the diurnal period, only one-half the dosage of nembutal was required to produce a latency of sleep onset equal to that observed after administering this drug in the nocturnal period (Levitt, 1963).

## B. *Number of episodes*

A second major aspect of the distribution of sleep behaviour across the 24 h is the number of sleep and waking episodes. As has been noted, the adult human organism typically displays a pattern of a single nocturnal sleep period and a single waking period from about 5 years of age. However, this pattern is clearly associated with age below 5 years in the human, and also shows considerable species differences.

As pointed out elsewhere (Webb, 1969) "The number of sleep periods, perhaps because of apparent theoretical barrenness, has received little attention in research" (p. 61). Parmalee (1961) reports the study of the sleep of 1 infant from birth to 8 months of age. In doing so, he carefully reviews the available studies from observations of children's sleep. His conclusions are as follows: "There is a striking similarity in the general

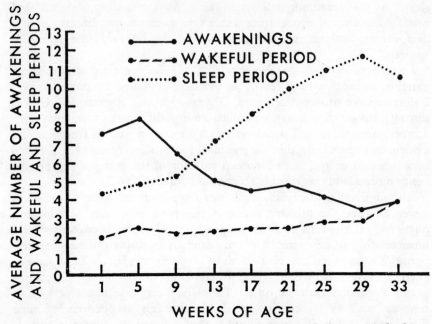

FIG. 9. Number of awakenings, wakeful periods and sleep periods for a single infant.

evolution of the patterns of sleep and wakefulness. In this case and the other single cases reported, as well as the larger group of studies previously mentioned ... essentially the changes in sleeping and wakefulness patterns from birth through at least the first 8 months are not so much a decrease in the total amount of sleep, but in the distribution of the periods of sleep and wakefulness ... this (sleep) is broken up into many short periods ... infants in the early weeks of life seem to be incapable of sustaining a long period of sleep or a very long period of wakefulness. As they mature, they acquire the ability to do both. Thus, the short periods of sleep consolidate into larger ones, especially in the night, and the short periods of wakefulness consolidate into larger ones, especially in the daytime." (p. 168).

In our own review of the available studies, we could find only 2 studies giving the number of episodes of waking or sleeping. The average number of awakenings reported by Parmalee on his single subject (per day frequencies of awakening based on weekly averages) as well as the average longest daily sleep and wake periods per week are given in Fig. 9. The further consolidation of these patterns is most clearly noted by Moore and Ucko (1957) who reported that the awakenings between midnight and 05.00 ceased for 70% of their sample of infants at the age of 3 months, 83% at 6 months, and 90% at age 1 year. Other relevant data was obtained by Reynolds and Mallay (1933), who reported on the decreasing role of the daytime nap. It constituted 12% of the sleep obtained between ages 1 and 2 and decreased to 5% by age 4. The nap had dropped entirely out of the sleep response of their nursery group by the age 5.

In summary, the pattern of the development of the uniphasic sleep response in the young adult appears to be a consolidation of a single sleep response into the nocturnal period with increasing lengthening of the wakeful period. This appears to be essentially achieved by about five years of age.

As shown in Table II, a clear cut pattern also emerges at the other end of the age continuum. Beyond the age of 30, we find increasing amounts of wakefulness and light sleep during a normal night's sleep. Feinberg and Carlson (1968) report the number of arousals as increasing linearly with age with an average of 6 awakenings in their population of subjects over 80 years old. These data are derived from EEG recordings. McGhie and Russell (1962) report similar results from a questionnaire study. In addition to this breakup of the uniphasic sleep pattern, a recent observational study of 40 elderly subjects (beyond 70) reports a mean number of 1·75 naps during the day (Webb and Swinburne, 1970).

As with the sleep of humans, the number of episodes of sleep in the circadian pattern in animals has received limited attention. The typical reports have been concerned with total duration and nocturnal–diurnal distribution.

## V. Envoie

In my introduction, I said that a psychologist venturing beyond his experimental data more often than not finds himself in deep shoals. I noted that for even the restricted consideration of circadian rhythms, conceptual problems manifest themselves with even the most cursory attempt to consider the meaning of the data relative to sleep. In another context, on the other hand (Webb, 1968), I attributed some of our limited progress in sleep research to the fact that ". . . data gathering in absence of theory or concepts may result in a mound of facts which often miss ultimate causes or effects or indeed may even bury them" (p. 56). Thus caught within my self-created dilemma, I shall gingerly put forth some highly speculative notions about my current beliefs concerning the function of sleep in the circadian process.

Stated as boldly and as simply as possible, my hypothesis (to dignify by word, my guess), is that sleep does not serve a "recovery" or "nutrative" function relative to waking behaviour, but rather serves a behaviour control function. The behaviour in this instance is that of non-responding. Stated otherwise, sleep is a process in aid of preventing non-adaptive, ineffective, or destructive behaviour from occurring rather than a time period in which recovery from behaviour episodes occurs.

There are 3 basic assumptions underlying this position:

1. Each species and/or individual within a species has a limited capacity to continue responding and a corollary requirement to not respond during a specified time period.

2. This non-response period is not randomly interjected into the circadian rhythm, but rather is adaptively keyed to an animal's capacity and environment.

3. Sleep is a critical part of this limitation of responding.

More narratively, I would suggest that each species and each individual within a species has an adaptive level of energy expenditure available to it in a given behaviour cycle of energy intake—environmental demands— energy expenditure. In man, for example, this energy expenditure period within the 24-h circadian cycle amounts to about 17 h (plus or minus high or low environmental demands); for a cow, this period amounts to about 21 h; for a rat, 12 h, etc. Relative to his energy resources, the animal must not respond during the remaining period of time or court disaster. Adaptively, when the internal body cues indicate the approach of a critical point of exhaustion and consequent damage or inefficiency (or both) these cues elicit non-responding. Furthermore, this non-responding adaptively occurs at the most appropriate time period within the circadian cycle so that its occurrence is least likely to result in danger from predators, or minimizes wasteful food gathering activities, or is most appropriate in terms of temperature, moisture and other environmental symbiotic events. In short, the non-response period is cued internally and is placed in time by environmental cues. Sleep serves the purpose of aiding or provoking this non-response event.

Let me elaborate this concept of sleep behaviour by argument and by some scarce data, since it runs contrary to at least 2 well embedded, but implicitly held ideas about sleep, namely, that sleep is a "non-adaptive period" and that sleep is a "recovery period".

A familiar puzzle in considering sleep which combines both of these notions goes like this: how can we account for the ubiquitous and continuing presence of sleep in normal behaviour of the animal kingdom throughout the evolutionary process when its "non-adaptive" aspects are so manifest? From this viewpoint, the organism is in, during sleep, a state that makes it least capable of coping with, adapting to, or protecting itself in the environment. Given such an evaluation of sleep, the answer has often been that sleep must indeed fulfill some very strong "need" to account for its persistence in spite of its apparent non-adaptiveness.

I am suggesting, to the contrary, that sleep and non-responding are in fact an adaptive form of behaviour the amount and placement of which are part of the process of behavioral adaptation. Within an environment, high responsiveness and energy expenditure are not necessarily adaptive, but may often be either dangerous or wasteful. Behavioural ecologists have emphasized that the behaviour of organisms is carefully geared through evolution to optimal time periods for food gathering, protection from natural predators, symbiotic temperature and moisture conditions, and the like. I am here suggesting that non-activity may be equally effectively geared to the environment when non-responding results in decreased dangers (e.g., remaining in a naturally protective environment) and decreased energy expenditure, when energy expenditure would be minimally effective. The simplest example that I may use would be primitive man. As a "day-active" and "night-inactive" animal, continued activity during the dark period would have increased danger and decreased effectiveness of performance. Night activity would have simply resulted in his falling off cliffs, drowning in bogs, being consumed by effective night predators—very poor returns relative to his energy expenditure. Such behaviour, in fact, would have resulted in his probable extinction.

Although I have recently read as widely as possible in the areas of behaviour ecology, I must admit that there is little direct support for my argument; whereas there is considerable and increasing data to establish the fact that peak reactivity is an adaptive process, almost no attention has been directed towards the adaptive qualities of non-responding.

In a most thoughtful paper "Survival Value of Diurnal Rhythms" Aschoff (1964) comes closest to expressing my concern, as follows: "Diurnal adaptation simply means doing the right thing at the right time, and thereby making use of the most favourable and **avoiding the unfavourable conditions** for survival" (p. 83). (Emphasis added) He adds: "The opportunistic nature of adaptation would not have achieved its goal if it had not used this (circadian rhythmicity) to increase the chances for survival" (p. 97).

I would cite a few observations that impress at least me. I have noted elsewhere (Webb, 1969) that although the distribution of sleep is moderately affected by light and dark, ". . . there is no intrinsic relation between darkness and sleep itself. A great number of animals sleep predominantly in the day and are awake in the night, but even within a given animal family, light and darkness have opposite effect in relation to sleep and waking tendencies. For example, in the rodent group, the cotton rat is, in general, a night sleeping animal, whereas the laboratory rat is, in general, a day sleeping animal (Van Twyver, 1967). This would suggest, of course, that light and dark are indexes of ecological conditions that favour the occurrence of sleep or wakefulness relative to the survival, convenience, or efficiency of the organism" (p. 58). Cloudsley-Thompson (1960) has noted at least 2 examples of the shifting of diurnal and nocturnal habits (and I would assume here associated dangers in peak responding periods) in an adaptive and short term adjustment. He notes that the African Buffaloes were almost eliminated in 1890 by an epidemic of Rinderpest. At that time the survivors, who used to feed in herds in the open by day, retreated to the forest and dense swamp and ate only by night. As the buffaloes increased in number, they returned to their former diurnal behaviour. He further notes that game animals tend to nocturnal habits when hunted extensively although they had previously been diurnal animals.

Even more deeply ingrained, although seldom explicitly stated, is the notion that sleep is a "recovery". In its simplest form, the idea is that the organism exhausts itself and then sleeps, during which time it "recovers". This recovery may take the form of reducing cumulative wastes or toxins or of rebuilding the energy supply for further use. The proposal which I have made places inactivity during sleep in the role of **exhaustion prevention** in contrast to **energy restoration**. Instead of sleep and inactivity being a direct "feedback" system, this position would view the energy depletion restoration system as a more long term averaging process in which there is a fixed component of inactivity associated with a species or an individual.

The primary data which would support such a conception would be evidenced in the lack of relationship between the amount and kind of waking activity and subsequent amount of sleep. I believe there is increasing provocative support for this position. I would first note that generally all species follow a basic 24-h period. Given such circumstances with biphasic events such as waking and sleeping, a long waking period automatically determines a short sleep period. By the nature of things, then, species differences and individual differences result in a negative correlation between waking and sleeping. Under such a circumstance, long wake periods followed by short sleep periods would represent limited recovery opportunities relative to the need developed. So far as I know, patterns of long waking and short sleep, or the reciprocal, short waking and long sleep, share only one established relationship to survival or behaviour efficiency, and in this case the relationship is a negative one. Recently, Zepelin and Rechtshaffen (1970) have reported a negative correlation between sleep length and longevity. Certainly the relationship between individual differences in length of **chronic** sleep length patterns and wellbeing in human subjects is not an obvious one. A recent study (Webb and Friel, 1970) comparing subjects who reported sleeping $5\frac{1}{2}$ h with subjects sleeping $9\frac{1}{2}$ h on intellectual, emotional and physical variables showed no differences. The conclusion of this study was as follows: "the data here gave no evidence that the extremes of these reported lengths are related to indices of mental, psychological, or physical status. This suggests, at least, that a maintained schedule of sleep which is self-selected as appropriate by an individual, does not appear to result in obvious consequences" (p. 66). Even more striking in this regard is the recent report of Jones and Oswald cited earlier (1968) describing 2 individuals whose sleep was limited to 3 h in 24. They found no obvious consequences of this short sleep "diet". Between species, there is an even more spectacular difference in length of sleep during the 24-h period. The possibility of following an "effectiveness" of behaviour or health dimension through species does not seem possible at this time. I would, however, simply point to the fact that ungulates (sheep, cows, goats, and the like) have been generally conceded as having a severe limitation on their "true sleep" and that this group of animals includes a broad variety of size, weight and reactivity. If the relationship between the amount of sleep and capacity to perform or survive were a simple one, one would certainly expect more uniform characteristics, both physiological and behavioural, in those animals showing limited sleep.

One piece of experimental data has "tested" the "feedback" model. Sleep of the rat is intermittent and widely varied across the 24 h in terms of length of sleep and waking episodes. In a demand situation in which waking episodes are essentially homogeneous in activity and stimulus input, one would expect long intervals of wakefulness to be followed by long intervals of "recovery" sleep. In fact, it has been found that the length of waking period in no way predicts length of subsequent sleep (Webb and Friedmann, 1969).

Within the "exhaustion–recovery" model, the effects of sleep deprivation are simply interpreted—they are a function of the omitted "recovery". How does the present "behaviour control" model handle this imposing problem? First, recognize that the proposed model does not deny "exhaustion"; however, we do say that the effects of deprivation are not due to the removal of a recovery period, but rather, replacement of non-behaviour with behaving. The effects of sleep deprivation, then, are not interpreted as a consequence of sleep (and recovery time) loss, but rather, are primarily due to the behaviour which occurs during the usual non-behaving time. This predicts that the effects of sleep deprivation will be primarily determined by the methods and provoked behaviour interjected into the non-sleep time. We look forward to testing this prediction.

Finally, let me point to a most critical aspect of this notion which requires explication. Namely, does sleep play a provocative or a supportive role in the non-responding that occurs? Does sleep as a process **cause** non-responding or simply aid in the maintenance of non-responding?

The most prominent advocate of an active inhibitor process was Pavlov, whose position is amusingly dismembered by Kleitman in his review of neurotheories of sleep (Kleitman, 1963). I would select Jacobson (1938) as the prime advocate of sleep as an adjunct to non-responding (in his formulation-relaxation). Recognizing the importance of this question, I have tried to arrive at my own conclusion. Space does not permit the complex of considerations that I have tried to take into account in this endeavour. I can only confess that at this point, I have not been able to satisfy myself with any degree of certainty. I suspect that the answers will emerge from the generally neglected field of sleep induction. I hope that the position I have stated will highlight this problem and that efforts towards its resolution will be stimulated.

In brief summary, I have suggested that sleep has developed as a behaviour control mechanism associated with reduced energy expenditure when that energy expenditure would have been dangerous or ineffective. The patterns of sleep and waking in turn have emerged from an adaptive interaction between the organism's physiological characteristics and capacities and environmental survival. Unfortunately the "experimental" variables underlying this proposition are embedded in the millions of years of evolution. It is my hope, however, that continued appraisal of the 24 h sleep and waking behaviour, particularly considering cross species and individual differences data, will permit an assessment of this proposition.

## References

Abt, L. E. and Riess, B. F. (eds) (1969). "Progress in Clinical Psychology, Vol. 8. Dreams and Dreaming". Grune and Stratton, New York.

Agnew, H. W., Jr. and Webb, W. B. (1968). *Psychophysiology* **5**, 142–148.

Agnew, H. W., Jr., Webb, W. B. and Williams, R. L. (1967). *Percept. Mot. Skills* **24**, 851–858.

Agnew, H. W., Jr., Webb, W. B. and Williams, R. L. (1968). *Psychophysiology* **5**, 216 (abstract).

Aschoff, J. (1964). *Symp. zool. Soc. Lond.* **13**, 79–98.

Aserinsky, E. and Kleitman, N. (1953). *Science* **118**, 273–274.

Baekeland, F. and Lasky, R. (1966). *Percept. Mot. Skills* **23**, 1203–1207.

Clemes, S. R. and Dement, W. C. (1967). *J. Nerv. Ment. Dis.* **144**, 485–491.

Cloudsley-Thompson, J. L. (1960). *Cold Spring Harbour Symp. Quant. Biol.* **25**, 345–356.

Dement, W. C. (1960). *Science* **131**, 1705–1707.

Dement, W. C. and Kleitman, N. (1957). *Electroencephalog. Clin. Neurophysiol.* **9**, 673–690.

Feinberg, I. (1969). *In* "Sleep: Physiology and Psychopathology" (A. Kales, ed.), pp. 39–52. Lippincott, New York.

Feinberg, I. and Carlson, V. R. (1968). *Arch. Gen. Psychiat.* **18,** 239–250.

Foulkes, W. D. (1966). "The Psychology of Sleep". Charles Scribner's Sons, New York.

Friedmann, J. K. (1969). "Contingencies of Sleep Deprivation". Unpublished Master's thesis, University of Florida.

Greenberg, R., Pearlman, C., Kawlische, S., Kantrowitz, J. and Fingar, R. (1968). "The Effects of Dream Deprivation". Paper presented at the March meeting of the Association for the Psychophysiological Study of Sleep, Denver, U.S.A..

Hartmann, E. L. (1967). "The Biology of Dreaming". Charles C. Thomas, Springfield, Illinois.

Hartmann, E. L. (1968). *Arch. Gen. Psychiat.* **18,** 280–286.

Hartmann, E. L. (1969). *In* "Sleep: Physiology and Psychopathology" (A. Kales, ed.), pp. 183–191. Lippincott, New York.

Hermann, H., Jouvet, M. and Klein, M. (1964). *C.R. Acad. Sci., Paris* **258,** 2175–2178.

Hobson, J. A. (1967). *Electroencephalog. Clin. Neurophysiol.* **22,** 113–121.

Jacobson, E. (1938). "Progressive Relaxation". University of Chicago Press, Chicago, Illinois.

Jones, H. S. and Oswald, I. (1968). *Electroencehpalog. Clin. Neurophysiol.* **24,** 378–380.

Jouvet, M. (1965). "Aspects Anatomo-fonctionnels de la Physiologie du Sommeil" (Lyon Symposium). Editions due Centre National de la Recherche Scientifique, Paris.

Kahn, E. and Fisher, C. (1969). *J. Nerv. Ment. Dis.* **148,** 477–494.

Kales, A. (ed.) (1969). "Sleep: Physiology and Psychopathology". Lippincott, New York.

Kales, A., Jacobson, A., Kales, J. D., Kun, T. and Weissbuch, R. (1967). *Psychon. Sci.* **7,** 67–68.

Kales, A., Wilson, T., Kales, J. D., Jacobson, A., Paulson, M. J., Kollar, E. and Walter, R. D. (1967). *J. Amer. Geriat. Soc.* **15,** 404–414.

Kety, S. S., Evarts, E. V. and Williams, H. L. (eds) (£967). "Sleep and Altered States of Consciousness". Williams and Wilkins, Baltimore.

Klein, M., Michel, F. and Jouvet, M. (1964). *C.R. Soc. Biol.* **158,** 99–103.

Kleitman, N. (1963). "Sleep and Wakefulness". University of Chicago Press, Chicago.

Kleitman, N. (1967). *In* "Sleep and Altered States of Consciousness" (S. S. Kety, E. V. Evarts and H. D. Williams, eds), pp. 30–38. Williams and Wilkins, Baltimore.

Kleitman, N. and Engelmann, T. G. (1953). *J. Appl. Physiol.* **6,** 269–282.

Koella, W. P. (1967). "Sleep, its Nature and Physiological Organization". Charles C. Thomas, Springfield, Illinois.

Kripke, D. F., Reite, M. L., Pegram, G. V., Stephens, L. M. and Lewis, O. F. (1968). *Electroencephalog. Clin. Neurophysiol.* **24,** 582–586.

Levitt, R. A. (1963). "The Experimental Control of Sleep Latency by Pentobarbital Sodium and the Influence of Visual and Auditory Stimuli on Sleep Latency". Unpublished Master's thesis, University of Florida.

Loomis, A. L., Harvey, E. N. and Hobart, G. (1935). *Science* **82**, 597–598.
Loomis, A. L., Harvey, E. N. and Hobart, G. (1937). *J. Exp. Psychol.* **21**, 127–144.
Luce, G. G. and Segal, J. (1966). "Sleep". Coward-McCann, New York.
McGhie, A. and Russell, S. M. (1962). *J. Ment. Sci.* **108**, 642–654.
Moore, T. and Ucko, L. E. (1957). *Arch. Dis. Childhood* **32**, 333–342.
Murray, E. J. (1965). "Sleep, Dreams, and Arousal". Appleton Century Crofts, New York.
O'Connor, A. L. (1964). "Questionnaire Response about Sleep". Unpublished Master's thesis, University of Florida.
Oswald, I. (1962). "Sleeping and Waking: Physiology and Psychology". Elsevier, New York.
Parmalee, A. H., Jr. (1961). *Acta Paediat. Scand.* **50**, 160–170.
Parmalee, A. H., Jr., Schultz, H. R. and Disbrow, M. A. (1961). *Pediat. J.* **58**, 241–250.
Pellet, J. and Beraud, G. (1967). *Physiol. Behav.* **2**, 131–137.
Rechtschaffen, A. and Kales, A. (eds) (1968). "A Manual of Standardized Terminology, Techniques, and Scoring System for Sleep Stages of Human Subjects". Public Health Service, Government Printing Office, Washington D.C.
Reynolds, M. M. and Mallay, H. (1933). *J. Genet. Psychol.* **43**, 322–351.
Roffwarg, H. P., Muzio, J. N. and Dement, W. C. (1966). *Science* **152**, 604–619.
Ruckebusch, Y. and Bost, J. (1962). *J. Physiol. (Paris)* **54**, 409–410.
Sampson, H. (1966). *J. Nerv. Ment. Dis.* **143**, 305–317.
Synder, F. (1964). "The REM State in a Living Fossil". Paper presented at the March meeting of the Association for the Psychophysiological Study of Sleep, Palo Alto, U.S.A.
Synder, F. (1969). *In* "Sleep: Physiology and Psychopathology" (A. Kales, ed.), pp. 266–280. Lippincott, New York.
Tauber, E. S., Roffwarg, H. P. and Weitzman, E. D. (1966). *Nature (London)* **212**, 1612–1613.
Terman, L. M. and Hocking, A. (1913). *J. Educ. Psychol.* **4**, 138–147.
Tremble, T. R. (1970). "The Effect of the Intensity of Presleep Visual Stimulation on REM Sleep". Unpublished Master's thesis, University of Florida.
Tune, G. S. (1969). *Brit. J. Med. Psychol.* **42**, 75–80.
Ursin, R. (1968). *Brain Res.* **11**, 347–356.
Van Twyver, H. B. (1967). "Analysis of Sleep Cycles in the Rodent". Unpublished doctoral dissertation, University of Florida.
Van Twyver, H. B. and Webb, W. B. (1968). *Psychophysiology* **4**, 364 (abstract).
Verdone, P. (1968). *Electroencephalog. Clin. Neurophysiol.* **24**, 417–423.
Webb, W. B. (1968). "Sleep: An Experimental Approach". Macmillan, New York.
Webb, W. B. (1969). *In* "Sleep: Physiology and Psychopathology" (A. Kales, ed.), pp. 53–65. Lippincott, New York.
Webb, W. B. and Agnew, H. W., Jr. (1962). *Science* **136**, 1122.
Webb, W. B. and Agnew, H. W., Jr. (1965). *Science* **150**, 1745–1747.
Webb, W. B. and Agnew, H. W., Jr. (1967). *J. Exp. Psychol.* **74**, 158–160.

Webb, W. B. and Agnew, H. W., Jr. (1969). *In* "Progress in Clinical Psychology, Vol. 8. Dreams and Dreaming" (L. E. Abt and B. F. Riess, eds), pp. 2–27. Grune and Stratton, New York.

Webb, W. B. and Friedmann, J. K. (1969). *Psychon. Sci.* **17,** 14–15.

Webb, W. B. and Friel, J. (1970). *Psychol. Rep.* **27,** 63–66.

Webb, W. B. and Stone, W. (1963). *Percept. Mot. Skills* **16,** 162.

Webb, W. B. and Swinburne, H. (1971). *Percept. Mot. Skills* **32,** 895–898.

Weitzman, E. D., Kripke, D. F., Goldmacher, D., McGregor, P. and Nogeire, C. (1970). *Arch. Neurol. (Chicago)* **22,** 483–489.

Weitzman, E. D., Kripke, D. F., Pollak, C. and Dominguez, J. (1965). *Arch. Neurol. (Chicago)* **12,** 463–467.

Williams, R. L., Agnew, H. W., Jr. and Webb, W. B. (1964). *Electroencephalog. Clin. Neurophysiol* . **17,** 376–381.

Williams, R. L., Agnew, H. W., Jr. and Webb, W. B. (1967). "Effects of Prolonged Stage Four and 1-REM Sleep Deprivation". U.S.A.F. School of Aerospace Medicine, Report SAM-TR-67-59, Brooks Air Force Base, Texas, U.S.A.

Witkin, H. A. and Lewis, H. B. (eds) (1967). "Experimental Studies in Dreaming". Random House, New York.

Wolstenholme, G. E. W. and O'Connor, M. (eds) (1961). "The Nature of Sleep". Little, Brown, Boston.

Wood, P. (1962). "Dreaming and Social Isolation". Unpublished doctoral dissertation, University of North Carolina

Wyatt, S. and Marriott, R. (1953). *Brit. J. Ind. Med.* **10,** 164–172.

Zarcone, V. and Dement, W. (1969). *In* "Sleep: Physiology and Psychopathology" (A. Kales, ed.), pp. 192–199. Lippincott, New York.

Zepelin, H. and Rechtschaffen, A. (1970). "Mammalian Sleep and Longevity". Paper presented at the March meeting of the Association for the Psychophysiological Study of Sleep, Santa Fé, U.S.A.

CHAPTER 5

# A Periodic Basis for Perception and Action

A. J. SANFORD

*Department of Psychology, The University, Dundee, Scotland*

## I. Introduction

The observation of periodicity in the electrical activity of the brain is now quite familiar. Probably the best known of all such periodic fluctuations is the alpha rhythm, which has a characteristic frequency of about 10 Hz. There have been attempts to correlate the measurable features of the alpha rhythm with a large number of variables—intelligence, psychomotor ability, and so forth. Here in this chapter we are concerned with possible connections between the alpha rhythm and **temporal** aspects of perceptual and motor performance only. We are concerned with alpha as a speed factor and as an atom of psychological time.

Throughout the first part of the chapter, the work showing relationships of these kinds is outlined together with some of the theoretical notions used by the experimenters to predict and explain their results. We then consider some work which adds to these results, but which has no overt reference to the alpha rhythm. This has been concerned with trying to examine the way in which the brain codes incoming information in time. The most commonly quoted view here is the "perceptual moment theory" —the theory that sensory data reach consciousness in discrete packages rather than in a continuous flow. Such a view, as we shall see, has been put forward as one way in which economical use could be made of the limited number of "components" of which the brain consists.

There are several issues tied up with the general question of a perceptual moment. For example, if everything falling within a discrete package is treated by the higher centres of the brain as **simultaneous**, then does the perception of time itself stand being broken down into "atoms" or moments? If so, then during the span of one of these moments there can be no cognizance of temporal order. In the next pages, we will look at all of these issues, and, as we will see, the purely psychological experiments allow an assessment of the various theories to be made with greater confidence than do those experiments which involve us in considerations of the alpha rhythm. However, there are some striking finds from the psycho-physiologist which must be accounted for. In the next section we will consider them.

## II. The Alpha Rhythm

When recording electrodes are placed on the scalp, oscillations in brain potential are recorded all of the time unless the subject is under deep anaesthetic. Over much of the scalp under appropriate conditions alpha band activity with a frequency of about 10 Hz and an amplitude of about 50 $\mu$V can be observed. The alpha rhythm is observed most clearly when the subject has his eyes closed and is thinking of nothing in particular. The rhythm is apparently suppressed during mental activity such as performing calculations, etc., and thus the observations of these gross electrical changes are associated usually with an inattentive state. There are wide individual differences in the generality of this statement, however, (see Walter, 1950).

The presence of this easily observable activity has led to much theoretical speculation. One suggestion is that each alpha cycle represents a single scan of some hypothetical mechanism which searches the stores of raw sensory data and transmits the information found to some more central part of the brain. Different accounts of such hypothetical mechanisms have been given by Pitts and McCulloch (1947), Wiener (1948) and Walter (1950). We will not describe the details of these views here—they are described very adequately in a recent review by Harter (1967a). It is fairly accurate to say that these scanning theories have one common assumption—that sensory inputs arriving within the duration of a single scan are treated as though they were presented **simultaneously** and all temporal order is lost. In other words they are allied to the Perceptual Moment Theory*—the theory that changes in the environment are per-

---

* "Perceptual Moment" is a phrase first used in this specific sense by Stroud (1955).

ceived through a series of temporarily discrete samples, the duration of which is called a "moment". It has been argued that this discrete sampling scheme is the most economical way in which the components of the brain activity in perception might be used (e.g. McReynolds, 1953; Harter, 1967a). If we accept this kind of argument it is not unreasonable that the onset of each moment might be associated with some observable neural activity which would appear approximately periodic in nature. Such a promise has been held out by the observation of the alpha rhythm.

It would of course not be correct to say that the only interest in the alpha rhythm has been in considering it a correlate of some hypothetical perceptual function; the clinical-diagnostic value of **electroencephalography** is probably the most important aspect of this whole field of research. Nevertheless as a relatively easily recordable feature of brain activity it is obviously of great interest to research psychologists as well as to neurologists. For this reason there is a great "natural history" value in the study of the relationship between features of alpha activity and well-known laboratory tests of psychological importance. As we have said, the main material to be reviewed here is that relating alpha activity to temporal aspects of performance.

Thus a distinction can be made between two general orientations in this field of research. The first is research stemming from the view that some aspect of the timebase of alpha activity represents the neural substratum of the (hypothetical) perceptual moment. To support this view it is necessary to demonstrate not only correlations between some temporal feature of alpha activity and perceptual behaviour, but also to show clearly that some form of the perceptual moment hypothesis **itself** is tenable. The second view, which is not so "strong" as a theory, is to suppose that alpha activity is sensibly related to behaviour and then to aks general questions as to why this is the case. For example a relation between alpha frequency and speed of motor activity may be demonstrated, but does not prove the moment theory. It may well be that both alpha frequency and motor activity are related to some other rather more general metabolic factor. The experimental literature, reviewed below contains experiments aimed at answering both general and specific questions.

Since alpha activity shows a reasonably good periodicity it is not surprising that it has been considered a "psychological clock" of some kind. In the work reviewed here the most commonly tested attributes of the activity have been its **frequency** on one hand and its **phase** on the other. We will consider each one in turn.

## A. *Studies related to alpha frequency*

Although alpha frequency is approximately 10 Hz it does vary some-what from subject to subject. If each alpha cycle **were** involved as a timing unit of some sort, then it might well be expected that subjects with fast alpha activity should be able to utilize a given number of temporal units in less physical time than those having slow alpha activity. Hence, on certain tests, they should be able to do a given amount of processing in less time. In general there has been a considerable amount of success in experiments within this kind of situation. Two kinds of tasks have received special emphasis—reaction time tasks and simple repetitive motor tasks.

1. *Sensori-motor tasks and frequency.* Mundy-Castle and Sugarman (1960) used a simple tapping task and found that speed when tapping as quickly as possible, as well as tapping at "natural" speed (free tapping) correlated significantly with the alpha frequency, comparing subjects.* On the basis of this Mundy-Castle concluded that alpha may well represent a varying state of central excitability relating to perceptual and motor speed, and that alpha might act as a "central timing mechanism regulating co-ordination of afferent with efferent signals". Other investigators were divided in relating alpha frequency to tapping speed, Rémand and Lesèvre (1957) being unsuccessful, certain others† being successful. In general it appears that the relationship between tapping rate and alpha is absent if **older** subjects are used (Mundy-Castle, 1962), which may mean that the central relationship (alpha with task) can only be shown if other more peripheral factors do not contaminate the picture. Heuse (1957) found that there was a positive relation between alpha frequency and tapping speed, as well as greater rapidity on other tests. A relationship between alpha frequency and speed of writing (natural speed) has been shown by Denier van der Gon and van Hinte (1959), who found a good correlation for psychiatric patients having normal alpha rhythms and writing normally. In this test the subject was required to write a test word ("momom") several times. The mean time for **writing the word**—not for the total operation— was taken as the indicator of writing speed.

The results certainly suggest that the alpha cycle might represent some sort of time unit in the organizational processes of perceptual-motor acts. Surwillo (1961) put forward just such a theory. He suggested that between-subject variations in reaction time might well be due to differences in

---

* Alpha was recorded when subjects were not working but simply sitting with eyes closed, a situation giving best alpha records.

† See Mundy-Castle and Sugarman (1960).

alpha period, if it takes a fairly constant number of units to organize a response to a stimulus. His experimental evidence is startlingly convincing, undoubtedly due to the great care he employed to use estimates of alpha frequency taken only when the signal was on, starting a sample reading at the first peak or trough after the stimulus had been presented and ending at the last peak or trough when the responses had occurred. He used a tone of 250 Hz with no warning signal, but a short and varied intersignal interval. Note that the signal was auditory—such signals gave no observable relationship in the **phase** experiments, a point requiring further investigation. Using 100 subjects he found a clear correlation between alpha frequency and simple reaction time (Surwillo, 1963).

Probably an even more striking point made by Surwillo concerns the relation of age to reaction time. It is well known that performance on reaction time tasks tends to be poorer with respect to speed in older subjects (see, for example, Welford, 1958). Surwillo had a range of ages from 28–99 years in his subjects, and found by partial correlation procedures that although reaction time was correlated with age, the correlation was due entirely to the fact that older subjects tend to have slower alpha rhythms. Indeed, when alpha period was partialled out there was a slight **positive** correlation between age and reaction speed! These data are of considerable importance not only in the present context, but also for the whole field of biological ageing, suggesting as they do that the whole effect of slowing with age is due to metabolic factors which are recognizable through alpha frequency. Thus for many purposes alpha frequency may be a more meaningful measure than chronological age.

Surwillo (1964a) went on to demonstrate that more complex **choice** reaction behaviour is related to alpha frequency in the same sort of way—there was once again a positive correlation for a sample of 54 subjects (34–92 years) between frequency and speed. The situation was that of the c-reaction: using tones of 250 Hz and 1000 Hz he required a response only to the higher frequency. Subjects with longer alpha periods made slower choice reactions than subjects with shorter alpha periods. Furthermore, the relationship between choice time and age disappeared when the alpha period was partialled out statistically, once more reinforcing the view that it is alpha frequency rather than chronological age which is the main variable underlying the age-reaction time relationship. Surwillo's results are shown in Fig. 1.

If alpha frequency really is an important variable in the speed of responding—and the evidence points strongly in this direction—then it is clearly desirable to manipulate its frequency experimentally with a view to observing any concomitant behavioural changes which might occur. For instance, if alpha frequency were to be increased by some experimental

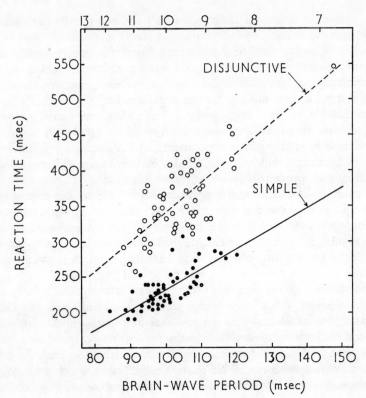

FIG. 1. Surwillo's results showing the relationship between reaction time and brain-wave period within the alpha band of frequencies. Reproduced with permission from Surwillo (1964a).

manipulation then it would be expected that reaction time should decrease. A result in this direction was obtained by Harter (1967b) who performed an experiment in which subjects breathed various concentrations of carbon dioxide. With concentrations giving a significant effect, reaction time and alpha frequency appeared to be in an inverse relation. It is clear however that such a finding is not very good evidence for any **direct** effect of alpha frequency on reaction time.

A better test was attempted by Surwillo (1964b) who used a technique known as "photic driving". Applying flickering visual stimulation at frequencies slightly above the natural frequency of alpha sometimes causes an increase in alpha frequency. Surwillo found that very few subjects in his sample manifested synchronization with the flash rate. Happily, there was some evidence of correlation between alpha frequency and reaction time in those subjects showing synchronization. This result, even if limited in some ways, is certainly consistent with the general picture. Suggestions for using other techniques such as sensory deprivation to modify alpha have been suggested (Obrist, 1965) but caution is obviously in order at this point: the use of techniques which act grossly on the organism may indeed affect alpha frequency and behavioural measures but this may be for **different reasons**. Suppose for example that the muscular state at the periphery were influenced by an experimental manipulation. It would clearly be far-fetched to argue that because alpha was influenced by the same gross manipulation that alpha was the causal factor in bringing about a change in reaction time. A far more satisfactory way of gaining greater insight into the potential role of alpha activity is to try to relate frequency and behaviour in a clear quantitative way. This is most plausible if **time** is used as a measure in both cases (e.g. reaction time and cycle duration) since the scale of measurement is clearly specified and the design and interpretation of experiments is correspondingly simpler. An example of this approach is discussed below.

2. *Kristofferson theory of attention.* Kristofferson (1965, 1966) put forward a theory of attention in which certain response parameters showed a striking correspondence with alpha frequency.     A digression is in order to explain certain features of his general line of thought. He postulates an axiomatic theory of attention which has as its core the notion of a single-channel processor. A key assumption he makes is that sensory projection pathways are divided into functional **channels** and that attention may be directed towards only one channel at a time. The situation is rather like that of a uniselector switch which takes only one input at a time. As an example the eye and the ear are supposed to be independent channels. He makes his theory even stricter by supposing that attention may be switched only at **discrete points in time** from one channel to another. Rather like the perceptual moment theory alluded to earlier, these switching points invite the possibility of looking for a physiological periodicity which might correlate with the switching points. The hypothesis he puts forward is of course extremely specific and it is difficult to see how such a specific hypothesis can be justified *a priori*. Nevertheless the results of some of his experiments are consistent with the view that alpha is related to perceptual-motor functioning.

His first experiments are concerned with the general properties of his model and are especially concerned with the estimation of the interval between switching times. An experiment of particular interest was concerned with psychometric functions generated in the task of perceiving the order in which 2 sensory events occurred. His theory predicts that the subject will correctly perceive the order of 2 events only if he can switch attention between the 2 events. If there is no opportunity to switch attention between the two then the subject will only operate at chance level. Clearly the probability of a switching time occurring during the interval between 2 events will depend upon the inter-event interval: it can easily be shown that the probability of a switch occurring is a linear function of the inter-event interval over the range of the switching interval. Thus the psychometric function relating probability of being correct to inter-event interval should be linear, and the event time difference required to change from chance performance to perfect performance should provide an estimate of the hypothetical switching interval.

On each trial of his experiment a standard light–sound pair was presented and a variable pair was presented. The terminations of light and sound were objectively simultaneous for the standard, but were asynchronous for the "variable" pair. The subject simply had to say which pair, the first or second, had asynchronous terminations. Customarily, ogives are used to describe psychophysical functions of the kind generated by this experiment and it is not easy to decide whether the linear or ogival curves give the best fit for Kristofferson's data. However, Kristofferson argued that there are good theoretical reasons for accepting the linear fit and continues on that assumption. The estimate for switching time obtained from the intercepts average about 55 msec, a value equivalent to about **half** an alpha cycle or about equal to the time it takes the cycle to cut the zero voltage line once. In a separate set of test sessions samples of alpha activity were taken. There was a positive correlation between alpha half-cycle time and the switching time estimates for the 8 individuals concerned, and the correlation reached statistical significance. Thus whether we accept the whole of Kristofferson's theory or not, we have additional evidence of the involvement of the biological rhythms of the brain in temporally-based performance. It is unfortunate that Kristofferson used only a very small sample of subjects in view of the fact that the reliability of alpha frequency measures is undoubtedly low, but even with a large sample the spread of alpha frequency across subjects would still be rather small.

In a later article (Kristofferson, 1967) it was pointed out that the linear fit used for the 8 practised observers in the earlier experiment does not always seem applicable. Sometimes unpractised subjects produced functions better fitted by **two** straight line segments (as shown in Fig. 2). The theoretical justification given to a 2-line function depends upon the argument that subjects do not always switch attention when a switching time occurs. Kristofferson supposes that any failure to switch attention from stimulus A to stimulus B will result in the need to wait for another switching point to occur. Such failures in switching may result in lack of practice or inattention to the task. It can still be argued that the psycho-

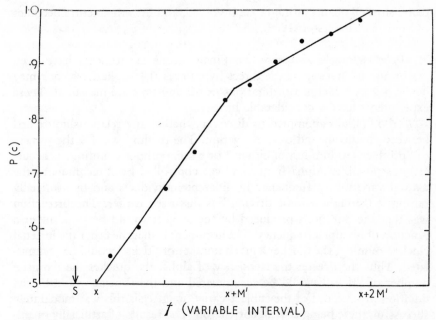

FIG. 2. Form of the psychometric function for successiveness discrimination under the 2-state mode of operation. Reproduced with permission from Kristofferson (1967).

physical functions relating the probability of a correct discrimination to the offset-disparity of the stimuli will be linear, but sometimes the functions will span two quanta (switching durations) instead of one, and the result will be the curve shown in Fig. 2.

By fitting 2 straight lines to data of the form of Fig. 2 for 13 subjects, Kristofferson obtained further estimates of the switching time constant in his model. EEG measures were made before and after each day's testing. Switching time estimates and half alpha cycle estimates (measured by independent observers) had the same mean value of 48 msec. Considering rank-order correlations between individuals, alpha half-cycle and switching-time estimates gave a correlation of 0·74.

These results are in agreement with those of Surwillo and the other investigators described in that they show a marked correspondence between alpha frequency and performance measures. They go somewhat beyond the earlier results in that they do show a fairly good quantitative correspondence between the intrinsic rhythm and the performance measure. In some of his earlier work, Kristofferson (1965) gives evidence suggesting that the same half-alpha time quantum is involved in reaction time under uncertain channel conditions, but in a later paper he suggested that a smaller quantum might be involved, and thus the picture becomes blurred (Kristofferson, 1969).

7. *Alpha cycle as a measuring rod*. Finally some experiments have been performed to investigate the direct hypothesis that alpha frequency may be a clock by which short durations of passing time are measured. These experiments have a considerable history.

Werboff (1962) attempted to discover whether any relationship existed between frequency and time estimation. The method was for the subject to "produce" a duration of either 2 or 8 sec, without counting, as accurately as possible. Alpha frequency was controlled by 2 techniques—the natural variation of frequency in different individuals and by artificially raising it through "photic driving" as discussed earlier. The prediction was that the durations produced by the subject would be some inverse function of his alpha frequency. The argument is that the faster the internal clock is running, the quicker a given number of "ticks" would be accumulated. Thus the greater the frequency of alpha, the quicker the requisite number of "ticks" would be accumulated and the shorter would be the duration produced. For the subject-variation analysis this was moderately successful, there being a weak correlation. The results of artificially speeding the alpha rhythm were, however, completely unsuccessful. A slight tendency to produce shorter durations with faster alpha rhythms was just significant for the 8 sec production only, but the explanation of this may well lie in the fact that "filled" intervals, that is intervals specified by observable exteroceptive stimuli, are judged to be longer than unfilled intervals of the same duration (e.g. Fraisse, 1964). The presence of a weak correlation when individuals are used as a source of frequency variation has been confirmed by Legg (1968) who required the production of a much wider range of intervals.

There is little support in these studies for the view that each alpha cycle is a "timing unit" used in duration estimation. Once again it must be stressed that general metabolic factors may be operating within individuals to cause both alpha activity changes and time estimation changes. There is some evidence to support this view. General body conditions such as age, thyroid state, thermic state and sensory deprivation are known to affect alpha frequency (e.g. Kleitman, 1963; Zubek and Welch, 1963). Such bodily conditions, especially those associated with changes in basal metabolic rate, are often associated with changes in the passage of subjective time (e.g. Hoagland, 1933; Baddeley, 1966). There is, therefore, a need to demonstrate for each task some kind of quantitative relationship between the particular performance measure being used and the frequency of alpha rhythm.

B. *Studies related to alpha phase*

A second line of attack in attempts to discover any psychological significance of alpha activity with respect to temporal aspects of behaviour has been made through studies of each alpha phase. For sine waves the definition of phase is perfectly simple and presents no difficulties. Alpha activity, however, is only very approximately sinusoidal and precision in deciding on the phase of any point on a record of alpha activity must inevitably be limited. Nevertheless there have been a considerable number of interesting observations made using phase as a variable.

1. *Reaction time and alpha phase.* The general design of this class of experiments is to present a stimulus to the subject and to have him react to it as quickly as possible. Monitoring the ongoing alpha activity enables a record to be made of that point in the cycle at which the stimulus was presented, from which an estimate of the phase can be made. By the same token, measurement of the phase at which the **response** was initiated can also be made.

Before describing the experiments the rationale behind some of the studies will be explained. Firstly there is the rationale that the alpha cycle is a fluctuation in the general excitability of the cortex. On this view we may well expect some sort of change in reaction time (as well as in say, subjective stimulus intensity) with the phase at which stimuli are presented. Secondly there is the class of ideas concerning the "perceptual moment" hypothesis that stimuli are sampled at discrete intervals of time. For this case a simple and strong prediction has been made by some workers. Longest reaction times will occur if a stimulus is presented when a new sample is just beginning, whilst the shortest reaction times should occur when a sample is just about to close. If, as has been suggested, the alpha cycle is related to the points where a sample opens and closes, then there should be a relation between the phase at which a stimulus is presented and the time it takes for the subject to recognize that it has been presented. Since the subject cannot react to the stimulus until he has perceived it, then reaction time should be a function of alpha phase. The maximum difference between reaction times obtained in this way should, of course, be equal to the duration of one "moment" or "scan".

An early investigation of this kind was performed by Walsh (1952) in the visual modality. He found no correlation between reaction time and alpha phase or amplitude. He also found no relationship between visual threshold and phase or amplitude. The later experiment of Lansing (1957) was more encouraging. He divided the alpha cycle into 6 arbitrary time divisions and found that the longest and shortest reaction times were indeed to light stimuli presented at opposite phases of the alpha cycle. However, between these 2 extremes the reaction times were not serially ordered at all, which tends to cast doubt on the whole enterprise. A further problem is that the difference between the longest and shortest reaction times was only of the order of 10 msec—a far cry from the predicted difference of 50 msec! This last fact means that the discrete sample or perceptual moment theory receives little or no support from Lansing.

Later researchers have demonstrated a more orderly correspondence between phase and reaction time. In one of the most recent studies (Dustman and Beck, 1965) a very **small** effect of phase on $RT$ was found— far less than the size of effect predicted by the perceptual moment theory. Experimental results obtained by Callaway and Yeager (1960) also showed a clear phase-reaction time relationship, but again the maximum difference was extremely small, being of the order of 10 msec. There certainly seem to be grounds for accepting that under some circumstances an effect of phase on $RT$ can be obtained, although it has not always been found (e.g. Walsh, 1952). Most recent studies aimed at finding this relationship have been successful, although certain complicating details do not seem to fit into any simple expected pattern. Firstly, the result has only been obtained for the visual modality. There are few reports of attempts to reproduce the effect in the auditory modality, and **no** successful reports (e.g. Callaway, 1962). The effect would thus appear to be restricted to the visual modality. Secondly, it has been demonstrated that the subjective brightness of a flash of light depends upon the alpha phase at which the flash was presented (Callaway, 1962). Such a result might be expected upon a simple "excitability-cycle" hypothesis and may appear to be in concordance with the reaction time results, since brighter stimuli are known to produce faster reactions (e.g. Roufs, 1963). However, such a simple-minded view is not particularly supported by the evidence, which showed that the phases associated with high brightness judgements were not related in any orderly manner to the phases associated with fast reaction times.

An intriguing attempt to discover at what stage of stimulus processing alpha phase exerts an influence on reaction time was carried out by Callaway (1962). There is some evidence that the well-known effect of brightness on simple visual reaction time may largely be mediated by changes in the transmission time of the retina (Bernhard, 1940; see also Vaughan, Costa and Gilden, 1966). Accordingly, manipulations of stimulus intensity should result in changes in the latency of the first arriving impulses in the brain. If the intervention of alpha takes place after the retinal stage, then, argued Callaway, the phase at which the stimulus is **presented** to give minimum reaction times should be different for bright and dim stimuli.

If we present a signal at a given phase of alpha, then by the time the neural impulses have reached the point where they interact with the alpha cycle the actual phase of alpha will have changed. For a weak signal the time for the impulses to reach this point will be longer than for a strong signal, and, accordingly, the phase at the interaction point will be different in the two cases. Thus, if a signal is presented at a certain phase to give a minimum reaction time for a dim signal, presenting a bright signal at this phase should not give a minimum reaction time because it should reach the point where it interacts with alpha too quickly.

Callaway chose the brightness of his stimuli to give a difference in reaction time of 50 msec, and anticipated that the difference in the phase of alpha associated with minimum reaction time should be 180°. The results, unfortunately, showed no effect of stimulus intensity on the phase giving minimum reaction time. Callaway's argument that alpha may affect reaction time so early in the course of neural events that there was no opportunity for the effect to show itself is difficult to accept if changes in latency due to intensity are indeed mediated by retinal processes. It is known that the effect of a change in **auditory** intensity does not give a fixed change in reaction time to the signal, but that the size of a loudness effect is dependent upon the adaptation state and attitudinal state of the subject (e.g. Grice and Hunter, 1964). Such findings strongly suggest a central locus for the auditory intensity effect. Not only that, but (again for audition) intensity does not markedly influence the perception of simultaneity in various psychophysical situations (Roufs, 1963; P. Bertelson, private communication; A. J. Sanford, unpublished data). These results show that the locus of the loudness effect in reaction time may not even be "sensory" in the customary use of the word. Although some results suggest that this may not be true for vision (Roufs, 1963), until we can define with certainty the locus of the intensity effect used in Callaway's experiment the meaning of his results cannot properly be appreciated.

Summarizing, although a small effect (*ca.* 10 msec) can be obtained by presenting a visual stimulus during different phases of alpha activity, the relationship to the psychophysical nature of the stimulus is confusing and poorly understood. A partial solution may lie at the motor end of the simple response act, although once again the details, as will be seen, are poorly understood.

2. *Response initiation related to alpha phase.* Reaction time, even simple reaction time, is the outcome of complex processing. Component processes cannot in general be directly observed but have to be inferred from experiments. It is possible to imagine reaction time as being the sum of delays from 2 kinds of components—predecision and postdecision. Certain

experiments suggest that the influence of alpha phase may be on the response initiation rather than on the predecision side, in contrast to Callaway's inference.

In one of the experiments described above, Lansing (1957) found a strong tendency for the motor response in simple reaction time to be emitted during a particular phase of the alpha cycle. A similar relationship had been reported previously by Bates (1951), who simply required his subjects to make a voluntary response, either a simple hand movement or an eyelid blink. When alpha activity was observable, there was a significant tendency for the initiation of movement to be related to the phase of the alpha cycle. The initiation of a movement was judged by the earliest signs of activity in the electromyograph—a device for measuring muscle potentials. Similarly, Kibbler and Richter (1950) claimed to find that a hand movement or an eyelid blink was related to the alpha cycle, in so far as the onset of muscular activity was not distributed in a random manner over all values of phase.

Recently a particularly thorough study of the role which alpha phase may play in determining the time of occurrence of a saccadic eye movement has been made by Gaarder, Koresko and Kropfl (1966). This represents an advance on the earlier work in that it uses the technique of computer-averaging of the EEG output—a method commonly used in recording evoked potentials in which successive records are summed so that only time-locked characteristics are added whilst "noise" is cancelled out. These authors made a record of eye movements and alpha activity whilst the subjects fixated a spot of light in a darkened room. Three different kinds of analyses were performed.

(a) Alpha activity was recorded on tape. This was then translated into a visual display from which it was possible to make a division of each alpha cycle into quadrants. The distribution of eye movements was not uniform over all quadrants, the effect being statistically significant.

(b) The saccades were used to trigger an averaging computer, so that samples of activity occurring just after each saccade were summed. Alpha-like activity appeared on summation. If the phase of alpha had been independent of the occurrence of a saccade, any alpha activity should of course have cancelled itself out in the averaging procedure. Indeed, random triggering of the computer resulted in an output displaying no alpha-like characteristics.

(c) The saccades were used to trigger the computer, but this time to sum activity **before** each saccade. The appearance of alpha-like activity strongly suggests that it is the **phase of alpha** which determines the time when a saccade will occur. Alpha activity is apparently not "reset" by the occurrence of a saccade.

Although the results of Gaarder *et al.* (1966), are not strictly applicable to the issue of simple reaction time, they are a step in the right direction. More important perhaps is the introduction of a technique (averaging) which might sensibly be applied to investigations of the time-locking of response production to alpha phase. Clearly, more demonstrations of response initiation-phase effects are called for, especially when it is appreciated that there are at least two features of the reaction time studies which go against the interpretation that it is through the time of response initiation that the alpha cycle interacts with simple reaction time. Such a view cannot possibly account for the fact that there is no apparent effect of alpha activity in the auditory modality. Worse than this, Callaway's attempt at obtaining a shift in the phase giving the minimum reaction time by varying stimulus intensity should have been successful. We are forced to conclude that although there are relationships between alpha phase and response initiation, alpha phase and reaction time, and phase and brightness, we do not understand the way in which these relationships interact.

What of the perceptual moment theory which was mentioned at the beginning of this section? The evidence reviewed above enables us to say very little one way or the other about its validity. It is true that the effect of phase on reaction time is only of the order of 10 msec, much less than the predicted maximum difference. If a full cycle (100 msec) represents a moment, then this would be the largest difference expected—and the result is 10 times smaller. However, even this result does not justify a rejection of the moment theory. Any source of error in the phases at which stimuli are presented will tend to reduce the difference between the longest and shortest reaction times if the moment theory is true. Since it is difficult to specify alpha phase with accuracy, and since alpha frequency and amplitude vary so that it is by no means truly sinusoidal, approximations to phase are the best which can be hoped for. Add to this the possibility of variation in sensory transmission time before the point where alpha may intervene, and the scale of the difficulties build up so that it would be a rash hope to really expect the minimum–maximum reaction time difference predicted by the perceptual moment theory even if the theory were true. Thus while the experimental results are interesting in a general way they are far from ideal for providing evidence with regard to the perceptual moment theory.

## C. *Comments on alpha*

Many dilemmas remain to be investigated. The phase-reaction time relationship has only been demonstrated for visual stimuli (and attempts at reproduction in the auditory modality have met with failure). On the other hand Surwillo's work with alpha frequency has succeeded with auditory stimuli and there are, to the writer's knowledge, no comparable data for the visual modality. This gap requires urgent attention, together with an attempt at understanding Callaway's results on different visual intensities in relation to alpha phase. Considering the fact that some investigators of simple reaction time are turning their attention to evoked cortical potentials and the "expectancy wave" observable during the foreperiod of a simple reaction task, perhaps attention will also be turned to the standard EEG records which often form part of the general records taken.

Perhaps the most important thing to do is to select situations giving the highest reliability for further examination. In this context Surwillo's results are the most encouraging. It would be interesting to learn to what extent his results hold for other latency situations, and to what extent the results are influenced by conventional variables such as number of stimulus alternatives, compatibility and so forth.

Finally, the technique of "Additive-Factors" described by Sternberg (1969) suggests a method of determining the different stages that go to make up a reaction time task. If this proves possible, it will open the way to answering the intriguing question of which particular parts of the sequence between stimulus and response are the ones influenced by alpha frequency.

## III. Psychological Experiments on the Moment and Periodicity

Stroud's (1955) conception of the moment was that sensory samples are taken at intervals defined by a periodic process, the exact period being possibly dependent on the nature of the task. When physiological data like alpha frequency are involved the periodic aspect is stressed even further. Logically, of course, perceptual moments need not necessarily always be of the same duration, but the general concordance of obtained estimates puts the size of the moment at about 100 msec (see later sections). We can ignore the invocation of physiological processes, and through purely psychological experiments endeavour to find evidence for the notion of a discrete moment, as Stroud himself has done. In what follows some of the areas in which the moment concept has been used are described, and in one case at least a direct test of the hypothesis has been carried out. The relevance of the hypothesis to the periodicity in performance is by now obvious—any repeated discrete process will be definable by a series of points in time which will have a mean value and thus imply periodicity. The principle decision to be made, therefore, is between a discrete or quantal model of perceptual input and a model which holds that sensory data are handled continuously. It will be obvious to the reader that different decisions may possibly be arrived at by considering different tasks. This section of the chapter describes some of the different tasks, but is not in any way exhaustive.

In addition, and particularly with reference to reaction time work, some sundry results are mentioned which imply a periodicity in performance probably unrelated to the main issue of the perceptual moment, but nevertheless of general interest.

## A. *Evidence from reaction time*

Reaction time distributions have been involved in attempts to test the perceptual moment hypothesis. Thus Stroud (1955) suggested that the delay between stimulation and perception would vary uniformly over a range equal to the span of a moment. The distribution of reaction time should therefore be rectangular with a width equal to the span of a moment and a standard deviation of one-third of the span of a moment. Stroud argues that such rectangularity does exist in some sets of data: "... the reaction times of single, very well trained subjects, obtained in single sessions, under very stable conditions, with foreperiods of appropriate lengths, form rectangular distributions ...". Unfortunately, it seems to be the case that reaction time distributions are far from rectangular. They are, in fact, usually positively skewed, and may well be described by a general gamma function (McGill and Gibbon, 1965).

Even if a moment type of process did underlie the production of reaction times it seems *a priori* absurd to look for rectangularity in their distribution since, as Stroud himself seems anxious to point out, under normal laboratory conditions this rectangularity is never seen. The weight of practical experience denies the hope of ever proving Stroud's prediction from the overall form of reaction time distributions, and all attempts to do this seem to have ceased.

On the other hand simple reaction time distributions do sometimes appear to be multimodal, as though some kind of periodic process underlies the probability of response occurring at any given time. Recently White and Harter (1969) have collected large numbers of simple reaction–time distributions from two subjects and found them to be multimodal in nature. They went on to plot the distribution of the intervals between successive peaks, and found that most of the peaks were separated by an interval of 25 msec. They suggested that this result was due to some periodic excitatory mechanism having a cycle-time of 25 msec. If electromyographic records are taken from the muscles involved in making simple reactions there seems to be evidence for a periodic fluctuation in voltage level also having a periodicity of 25 msec. Harter and White took such readings under both resting and reaction conditions and again analysed the distribution of intervals between successive peaks. The same subjects were used as were considered in the reaction time study. The outcome was that the distributions of between peak intervals were the same (or very similar) for both reaction time and EMG records. Similar signs of a periodicity of exactly the same order of magnitude have been found by Latour (1967) for the reaction time of eye movements and the detection state of the eye.

Quite clearly these signs of a very rapid periodic influence on performance cannot be allied to the suggestion of a psychological moment having a periodicity four times as long. Neither can they be thought of as the same as the kind of results relating alpha frequency and phase to performance. Rather they suggest that many, possibly independent, periodic mechanisms may influence behaviour.

A different kind of periodicity in reaction time distributions has been reported by Venables (1960). He used a simple verbal reaction to a circular visual patch and obtained 12 distributions of 200 reactions, 6 from normals and 6 from schizophrenics. When the data from the 6 normals and 6 schizophrenics were summed (separately) there was evidence of multi-modality with peaks 100 msec apart. Such a finding can be taken as evidence for a periodic tendency for a reaction to be elicited. Unfortunately, however, it cannot be related in any way to the attempts to show that reaction time is dependent on the phase of an ongoing rhythm such as alpha activity. The signal to react would presumably arise in random relation to an ongoing central cyclical mechanism and so should not manifest itself as a periodic fluctuation in response tendency revealed by multimodality. Venables, in fact, put forward the much simpler and much more plausible notion that the answer may lie in the form of the cortical evoked potential. In response to a stimulus a series of waves, sometimes evenly spaced (Chang, 1950), follows the primary response. Venables tentatively suggested that if the reaction was not initiated by the first wave, then it would be initiated by the second or third, and so on.

This is obviously a somewhat *ad hoc* speculation but it does make the point clear that in order to find periodicity in reaction time distributions the assumed underlying process would have to be time-locked to some aspect of the stimulus. Further attempts to understand effects such as those of Venables must await a better understanding of any interactions between ongoing and stimulus-locked periodic activity.

In conclusion, work on reaction time has not led to any sensible conclusion with respect to the psychological moment theory. The rectangularity in $RT$ distributions predicted by Stroud (1955) is seldom, if ever, really observable. There is some evidence for multimodality in simple reaction time distributions, though the kinds which we have examined must clearly be the result of different underlying periodicities. Finally, the biggest practical objection to any attempt to find periodicity is that it depends very much on the assumption that the periodicity in question has a somewhat limited range of variation and is more or less constant within a session.

We now go on to examine a second kind of experimental situation with respect to which the perceptual moment theory has been invoked—that of phenomenal simultaneity.

### B. *The perception of order and simultaneity*

1. *Two stimuli.* On the most popular version of the discrete perceptual moment theory failure on the part of the subject to correctly discriminate the order of 2 events is explained in the following fashion. If the stimuli are sufficiently close together, then both will have a certain likelihood of falling into the same perceptual moment. Accordingly, when the subject becomes aware of the stimuli he will be aware of them as **simultaneous** because stimuli encoded in the same perceptual sample yield no order information at all. By the same token if stimuli fall into different perceptual moments then they will be perceived as successive and the subject should, in theory, be able to say in what order the events occurred. Clearly, the probability of 2 stimuli falling into the same moment will depend on the interval between the stimuli, and provided the duration of a moment is constant the psychophysical function relating the probability of a correct order judgement to the interval between the critical stimulus events will be linear, just as was described in the earlier section on Kristofferson's work. In describing these results, however, it was pointed out that it is, in practice, very difficult to decide on the basis of the points on psychophysical functions between linear and ogival fits to the data.

Kristofferson's experiments, it will be recalled, yielded a value for the successiveness threshold which is rather small, the whole function from chance to perfect performance spanning only 50 msec, which, taken as a value for the perceptual moment would be half the figure commonly suggested as the sampling rate.

Other investigators have found comparably small values for thresholds using more conventional paradigms. Thus Hirsh (1959) required subjects to state the order of 2 different sounds presented in close contiguity. In order for judgements to be correct 75% of the time an interval of only 20 msec between the presentation of the 2 stimuli was necessary. The interval required for very high accuracy (95%) was less than 100 msee. An objection to this experiment was raised by Broadbent and Ladefoged (1959). They suggested that Hirsh's subjects may have been able to judge the order of the events on the basis of differences in quality between stimulus pairs presented in each order. In other words, the stimulus complex AB may sound different from BA so that after adequate practice such qualitative cues could be used to make the necessary judgement. Indeed, Broadbent and Ladefoged found that unpractised subjects were still performing at chance level although the separation between the stimuli exceeded 150 msec.

Hirsh and Sherrick (1961) countered this objection somewhat by showing that the same threshold values held for visual and tactual modalities as well—and that the same results could be obtained when stimulus pairs were made up of items from different sense modalities. The argument is that qualitative interactions are less likely in this case, although this is extremely difficult to prove.

Although Hirsh and Sherrick's finding that thresholds of order perception are the same for the major modalities indicates that there may be a central limitation at work, beyond this we cannot go. Obviously the invocation of an idea as complex as that of a discrete perceptual moment is in violation of the principle of parsimony. Efron (1967) has suggested that the perception of order is most likely a simple function of the interval between the first and second stimulus. This basic idea is that it takes a certain "processing time" for a stimulus to be perceived. If a second stimulus is presented quickly enough after the first stimulus, then the critical attributes of the second stimulus will be incorporated into the first stimulus precept and they will seem to be simultaneous. In this instance it is the interval between the onsets of the first and second stimulus which is the crucial variable.

The perceptual moment view point is somewhat different. In this case the critical variable is **not** the gap between the 2 stimulus onsets but is the **total time** in which the 2 stimuli fall **including** the interstimulus interval. An attempt to assess the relative importance of interstimulus interval and total presentation time has been made by Thor and Spitz (1968). They presented 2 triangles ($\triangle$ and $\triangledown$) successively at the same spatial locus and required the subjects to say in which order the events occurred. By varying stimulus duration and interstimulus interval (durations 10, 30, 50, 70 and 90 msec—always the same for both stimuli) and total time between 30 and 270 msec where possible, they were able to determine whether total duration or interstimulus interval was the critical event involved in the accuracy of discrimination. The results supported the total time concept, that is, accuracy was dependent upon the total time rather than on the interval between the 2 stimulus onsets. This total time had to be 100 to 120 msec for perfect accuracy to ensue.

2. *Repetitive displays.* Further evidence for this point of view comes from a different type of experiment using repeated displays. Lichtenstein (1961) used a display consisting of 4 very small light sources arranged in a diamond pattern and subtending a visual angle of 1°. In the experiment each dot was lit very briefly in succession so that the sequence 1, 2, 3, 4, was repeated over and over again. When the cycle rate was sufficiently rapid the appearance of the dots seemed simultaneous—i.e. they flickered in apparent synchrony. By manipulating the within-cycle temporal separation of each flash, a test could be made of the hypothesis that the real inter-stimulus interval is not important in determining discriminability.

The subject's task was to increase the repetition rate until the dots appeared to flash simultaneously, using the bracketing procedure of the method of adjustment. When all the dots were regularly spaced the total cycle time was 125 msec for simultaneity. The next stage was a manipulation of the interval between stimuli within the cycle. Even when five-eighths of the total cycle was interposed between 2 successive dots, at a cycle time of 125 msec there was no change in the threshold of simultaneity! Lichtenstein too felt confident that if the entire cycle fell within the span of one moment then there would be no effect of interstimulus interval.

Using a similar situation Allport (1966, 1968) performed some very interesting tests of the perceptual moment theory. Allport argues that there is no evidence to suggest that the span of simultaneity is quantized discretely in time, but that a continuous "time averaging" process is, *a priori*, equally possible. This hypothesis is termed by Allport the "Travelling Moment" theory. If we imagine a series of inputs at different times $t_1$, $t_2$, $t_3$, $t_4$, $t_5$ ... $t_n$, then on the discontinuous model a sample may contain firstly inputs at $t_1$, $t_2$, and $t_3$, then inputs at $t_4$, $t_5$, $t_6$, etc. On the travelling moment model firstly inputs $t_1$, $t_2$, and $t_3$ are sampled, then $t_2$, $t_3$, and $t_4$, then $t_3$, $t_4$, $t_5$, etc. Inputs sampled on both schemes are assumed to appear simultaneous if they fall into the span of the moment, though one scheme is discrete and possibly periodic, whilst the other is effectively continuous in operation. Allport's experiments decide between the discrete and continuous models.

His display consisted of a horizontal line on an oscilloscope which jumped intermittently to a new position a little above or below the previous one. Displays were always ascending or descending, the lines never appeared at random points. At faster rates the subject saw an increasing number of lines in apparent movement up or down the screen.

As the number of lines seen became greater, Allport reported a figure-ground reversal so that the display looked like a series of stationary lines with a dark band moving across it. At faster repetition rates still, the display appeared as a series of stationary lines. The point at which this occurred Allport called the point of simultaneity.

Several quite elegant predictions were made regarding the point of simultaneity under conditions of increasing or decreasing the repetition rate, but will not be discussed here because the most striking result came from the simplest observation. This observation concerned the direction of motion of the dark band (or "shadow") which appeared to move across the lines on the oscilloscope at repetition rates just less than that necessary for simultaneity to be reached. Allport argued that this "shadow" was due to one of the 12 line positions on the oscilloscope falling out of the reach of a moment. Consider the discrete theory. As the lines are presented, in order, from 1 to 12, suppose one of them to be excluded from the sample—say number 12. The next sample will include 12 but exclude 11, and the next will include 11 but exclude 10. Thus, on the discrete theory, the excluded lines will be a descending sequence whilst the display is, of course, ascending. In other words, the shadow will move in a direction opposite to that in which the lines are presented.

For the travelling moment the shadow should move in the **same** direction as the display direction rather than opposite. This is easy to demonstrate by considering lines 1 to 12 again. Suppose 12 to be beyond the reach of the span, which stretches over 1 to 11. The next stage will include 2 to 12 but exclude 1, and the next stage will include 3 to 1 but exclude 2. Thus the omitted lines are moving in the **same** direction as the display lines are occurring. The results were unequivocal. The shadow always moved in the same direction as the order of illumination of the lines. This result is quite incompatible with the perceptual moment theory in its discrete form, as postulated by Stroud.

We can summarize by saying that on the discrete perceptual moment viewpoint it is the chance of 2 or more events falling within one moment which determines the chance of making a correct response. Two experiments tend to support the view that it is total time and not the interval between the stimuli which is the important variable in successiveness. On the other hand, Allport's results cannot be explained on the discrete moment theory and he has put forward an alternative formulation as the "Travelling Moment Theory". Also of interest is Allport's observation that the span of simultaneity can be reduced by an increase in the intensity of the display, so that any invocation of a periodic process would have to allow for a simple change in frequency with intensity.

There are clearly many theoretical details to be worked out, but the most important single thing to be looked into must be differences in the **size** of the simultaneity threshold obtained by simple procedural differences. For instance, the perceptual moment viewpoint makes no distinction between the judgement of synchrony (e.g. Lichtenstein, 1961) and the discrimination of order. It may well be that this is not so, and indeed Murphree (1954), using a variety of tasks of temporal discrimination in the repetitive display paradigm, obtained results which pointed to different thresholds (in terms of cycle time) for the various tasks involved. There is no simple answer to this from the perceptual moment theory.

## C. Temporal numerosity and response rate

1. *A description of the phenomenon.* A rather different approach somewhat related to the issue of phenomenal simultaneity has been used by White and his co-workers (Cheatham and White, 1952, 1954; White, Cheatham and Armington, 1953; White, 1963). This, like the simultaneity experiments, provides a means of investigating the temporal resolving power of the human senses. The basic paradigm is deceptively simple. Rapid clicks or flashes are presented to the subjects at rates of 10, 15 or 30 items/sec. The task of the subject is to report how many items they thought they saw or heard. The actual number of items is varied as well as the rate of presentation. Under circumstances such as these it has long been known that subjects make a very special and reliable type of error: they always **underestimate** the number of events which are presented. The reliability of this phenomenon is such that the basic experiment has borne many replications (for vision: Forsyth and Chapanis, 1958; for audition; Garner, 1951; A.J. Sanford—unpublished data; in addition to those already cited). The phenomenon has been called by Cheatham and White (1952) "temporal numerosity" and it has an important practical application in establishing the upper limits of morse-code reception rate.

A typical set of data, derived from an auditory version of the experiment, is shown in Fig. 3. Very similar results have been obtained from both the visual and tactile modalities. Clearly the universal nature of the tendency to underestimate is strong evidence that subjects are not just guessing but that some potent mechanism is at work restricting the number of events perceived. The most interesting feature of the results is that the number of items reported seems to be a function of the duration of the stimulus train as a whole rather than a function of the actual number of items per unit time. Thus for rates of 10, 15, and 30 events/sec the subject reports about 10 items as being presented in 1 sec of total stimulation. There is a slight tendency for the perceived number to increase with rate, but the increase is small compared to the increase with time. For low rates

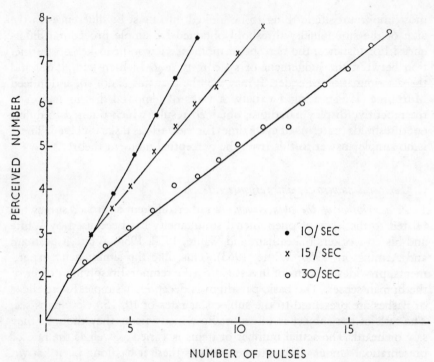

NUMBER OF PULSES

FIG. 3. Perceived versus presented number of pulses illustrating the auditory numerosity function. In this case, the mean is used as a measure of central tendency. Reproduced with permission from Cheatham and White (1954).

of stimulation there is no tendency towards underestimation and resolution is perfect.

The interpretation of these results is difficult. At sufficiently high rates of presentation subjects must tend to perceive as one stimulus all stimuli falling within a time period of 100 msec. White's (1963) assumption is that however many stimuli fall into one perceptual "sample" or moment only one stimulus will be perceived, since order information is lost within a moment and there are no other clues as to how many items have been presented. This argument is assumed to hold for all sensory modalities.

If there is such a central limitation then it should act on all sense modalities equally, hence the results from different modalities should be identical. In fact for vision if the duration of the stimulus train is shorter than about 300 msec then there is a tendency to report less stimuli as having been presented than in the corresponding auditory case. Beyond 300 msec the rate of gain of perceived items becomes first similar, then the same for the 2 conditions. If the mode is used as a measure of central tendency (White, 1963), this rate is 6–7 items/sec. The explanation of the initial discrepancy appears to lie in a peripheral complication of the visual case which is present only for the first initial period and which White has called the "initial fusion period". In a small series of experiments White (1963) gave evidence to show that the initial period was dependent upon the exact conditions of stimulation whilst the effect of stimulation beyond 300 msec was not dependent. In this way White argues that the 6 or 7 items/sec slope obtained after the initial period is the outcome of a truly central limiting process.

Physiological data provided by White and Harter (1969) supports the standpoint that the limitation is indeed perceptual. Thus the number of major peaks in the visual evoked cortical potential corresponded exactly to the modal number of reported flashes, although this time they found that one perceived item and each major peak occurred at a separation of about 100 msec. However the total stimulus train duration was not really long enough to afford a full comparison with earlier results. Other results have shown that if the limitation is perceptual it is not peripheral in origin. A further study quoted by White (1963) showed that the eye was quite capable of following rates of stimulation as rapid as 30 stimuli/sec. Electroretinographic (ERG) recordings were made under the same conditions as the numerosity study and the results dispelled the view that the phenomenon was due to an inability of the eye to respond with sufficient rapidity.

2. *Problems in numerosity work.* One is tempted, like Stroud (1955) to suppose that the numerosity data provide evidence for the existence of a discrete perceptual moment. Certain facts point to a very great limitation to the generality of any simple way of viewing these findings, however. For example, one would expect the number of items perceived in the numerosity situation to be the same as the number of items perceived in a situation where the primary task was to evaluate the subjective **rate** of a visually flickering stimulus. In each case the limiting variable should be the number of items perceived—in the flicker judgement situation subjective rate should be the number of items perceived per unit time. However, apparent rate of flicker is dependent upon the locus of stimulation with respect to the fovea—apparent rate being lower at the periphery than at the centre (Le Grand, 1937). This is not true of temporal numerosity in which the number of reported events is not influenced to any great extent by retinal displacement (White, 1963; Forsyth and Chapanis, 1958). The conclusion must be that the 2 mechanisms are independent, which rather spoils the picture. Figure 4 illustrates data from Lichtenstein *et al.*

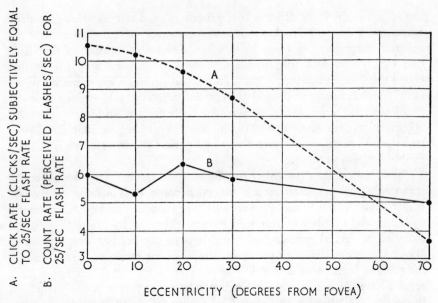

FIG. 4. Matching click rate to flash rate (dotted lines) yields a function dependent on the eccentricity of stimulation. The apparent rate of perception deduced from temporal numerosity studies yields a function only very slightly dependent on eccentricity (solid line). Reproduced with permission from Lichtenstein *et al.* (1963).

(1963). The dotted line is the apparent flicker rate, an estimate of which was obtained by requiring subjects to match a variable auditory click rate to the flicker. The solid line shows that the numerosity "rate" is more or less constant whilst the subjective flicker rate (dotted line) is heavily dependent on retinal displacement. The discrepancy has not yet been explained, but it may be that short trains of flashes would give different apparent flicker values.

There are other complications in the auditory modality. Although the subjective rate from numerosity studies is similar in audition to vision, fluttering sound does not sound fused until very high rates indeed have been exceeded and below this subjects can easily perceive the difference between various interruption rates. In other words, although the subjective rate derived from numerosity experiments is hardly influenced by the objective rate of stimulation, subjects can, nevertheless, tell that the rates are different. This fact, together with the visual flicker problem, makes it difficult to accept a perceptual explanation of the phenomenon.

In any psychophysical task there must be adequate provision for the subject to be able to make a response appropriate to the stimulation. Is this condition met by the numerosity paradigm? Suppose that the maximum "perceptual" rate was 20 items/sec, then in order to enumerate them accurately the subject would have to be able to count faster than 20 items/sec. However, many simple repetitive acts seem to have an upper limit much lower than this. Thus it is impossible to tap at rates faster than about 8–10/sec (e.g. Seashore, 1938), although by means of electrical stimulation of the muscle much faster contraction rates are possible. Similarly, syllabification rates are in the same vicinity (Miller, 1951). It is a simple matter to demonstrate that it is impossible to vocalize digits from 1 to 10, even without regard to pronunciation, at rates exceeding 8–10 items/sec. This rate seems to hold for other tasks involving very rapid counting (Saltzman and Garner, 1948).

If the rate of counting overtly is limited to such an extent then the question arises as to how rapidly subjects can count covertly, as they do in the numerosity stituation. It seems quite likely that rates of internal and external counting and speaking may be much the same (Landauer, 1962) and if this is the case then there will clearly be a severe response limitation on the Numerosity task, especially when the presented number of items exceeds a very few items which may be "counted" in parallel, rather like the span of apprehension for dots situation. This criticism might be quite important for White's analysis which relies so heavily on the late part of the numerosity function. A response limitation of this kind could quite well explain the failure of accuracy but as it stands does not explain the tendency towards **underestimation** which is, of course, the most reliable fact to be gleaned from the whole enterprise.

There is no easy way to explain this underestimation phenomenon, but the answer might well lie in the certainty with which the subject makes his responses. When a series of 8 tones at a rate of 30/sec is presented to the subjects they are not very confident of their answers. Perhaps all the subject can do is to "count as quickly as possible" and give the result as the most confident answer possible. This will inevitably cause underestimation and will result in the answers being more tied to the stimulus duration than to the actual number of stimuli presented. This objection cannot be made strongly, of course, since underestimation does occur when only very few items are presented at the beginning of the sequence. Nevertheless, it does indicate that more attention should be given to the early rather than the late part of the numerosity function. It does carry one testable implication: if subjects are really so uncertain (which they seem to be from personal observation) then it should be possible to obtain an apparently faster "perceptual rate" by giving subjects a second-choice response in addition to their first choice, so that they get a chance to express their uncertainty.

Come what may, attempts must now be made to explore the **nature** of the underestimation normally obtained with a view to discovering its locus. Only when this has been done can the numerosity results be sensibly described and seen in relation to other views of the perceptual moment. As it is the obtained results suggest a maximum "perceptual rate" of between 6 and 10 items/sec, which is in the region obtained in other experiments investigating the moment. The relation of a hypothetical "perceptual rate" to physiological brain rhythms remains in the realm of speculation. Certainly rhythmic fluctuations with a periodicity in the range implied by the Numerosity work may be found in cortical potential, both stimulus-locked (e.g. Chang, 1950) and unsynchronized, but any causal relation remains unproven.

## IV. Further Comments and Conclusions

The evidence reviewed demonstrates that in some circumstances a relationship between observable alpha activity and performance is a reality. The strongest relationship appears to be that between alpha period and reaction time. Work related to alpha phase is less satisfactory, not because of a total failure to demonstrate a relationship, but because the effects obtained are so small and are subject to day by day variation and because certain simple predictions have not been borne out in practice. Attempts to account for reaction time-phase relationships are not, as they stand, explicable in terms of a "perceptual moment" theory such as that of Stroud. The correlations of alpha frequency with reaction time have been used to postulate something analogous to a computer "cycle time" in the human brain, but there has been no real proof of this; it remains an intriguing possibility.

A variant on the perceptual moment theory is Kristofferson's discrete switching point model of attention. Once again correlation of the derived estimates of the switching intervals with alpha period (half-cycle in this case) was demonstrated. How this relates to Surwillo's findings on reaction time remains to be established. On the purely behavioural side there seems to be little evidence **directly** supporting the idea of a discrete sampling mechanism such as that which Stroud favoured, although the results on counting rapid sequences of flashes or tones remain difficult to account for in any other way. Some of the findings from the field of phenomenal simultaneity point strongly to a non-discrete "travelling moment" system which, since it is continuous, is not periodic. There is thus evidence **opposed** to the discrete moment theory from this front.

Turning to reaction time, there is evidence that periodic mechanisms underlie the sometimes observed multimodality in reaction time, although once again this evidence does not relate in any obvious way to the perceptual moment theory. It does, however, demonstrate that observable fluctuations in muscle potentials are intimately related to reaction time, and provides another example of how biological peridocity influences performance.

As was pointed out earlier, the evidence reviewed here is but a small portion of the whole range of investigations which have been performed. In general, there has been a trend away from any attempt to make a grand but parsimonious synthesis of all the facts of temporal discrimination. Rather, specific phenomena are being handled in specific ways but often with some connection, remote or close, to Stroud's ideas; for example Kahneman (1968) for visual masking and Shallice (1967) for the temporal integration of brightness. There is little doubt that such trends are healthier than attempting a grand synthesis at this stage of knowledge.

It appears from the simultaneity data at least that the discrete moment formulation is quite incorrect for this particular situation. Is this completely incompatible with the results of Surwillo and the temporal numerosity studies? The answer must be negative. It may well be that discrete time processes do exist but that rather than acting at what might be called the "perceptual" level they operate at some other level—for instance in sensori-motor integration. Until the problem has been tackled more analytically in each particular situation there can be no definite conclusion regarding the value and validity of the idea of fixed, discrete processing periods.

## References

Allport, D. A. (1966). Unpublished Ph.D. Thesis, University of Cambridge.
Allport, D. A. (1968). *Brit. J. Psychol.* **58**, 395–406.
Baddeley, A. D. (1966). *Amer. J. Psychol.*, **79**, 475–479.
Bates, J. A. (1951). *J. Physiol. (London)* **113**, 240–257.
Bernhard, C. G. (1940). *Acta Physiol. Scand.* **1**, Supplement 1.
Broadbent, D. E. and Ladefoged, P. (1959). *J. Acoust. Soc. Amer.* **31**, 1539.
Callaway, E. III. (1962). *Electroencephalog. Clin. Neurophysiol.* **14**, 674–682.
Callaway, E. III and Yeager, C. L. (1960). *Science* **132**, 1765–1766.
Chang, H. T. (1950). *J. Neurophysiol.* **13**, 235–258.
Cheatham, P. G. and White, C. T. (1952). *J. Exp. Psychol.* **44**, 447–451.
Cheatham, P. G. and White, C. T. (1954). *J. Exp. Psychol.* **47**, 447–451.
Denier van der Gon, J. J. and van Hinte, N. (1959). *Electroencephalog. Clin. Neurophysiol.* **13**, 923–926.
Dustman, R. E. and Beck, E. L. (1965). *Electroencephalog. Clin. Neurophysiol.* **18**, 433–440.
Efron, R. (1967). *Ann. N.Y. Acad. Sci.* **138**, 713–729.
Forsyth, D. M. and Chapanis, A. (1958). *J. Exp. Psychol.* **56**, 385–391.
Fraisse, P. (1964). "The Psychology of Time". Eyre and Spottiswoode, London.
Gaarder, K., Koresko, R. and Kropfl, W. (1966). *Electroencephalog. Clin. Neurophysiol.* **21**, 544–551.

Garner, W. R. (1951). *J. Exp. Psychol.* **41**, 310–316.
Grice, G. R. and Hunter, J. J. (1964). *Psychol. Rev.* **31**, 247–256.
Harter, M. R. (1967a). *Psychol. Bull.* **68**, 47–58.
Harter, M. R. (1967b). *Electroencephalog. Clin. Neurophysiol.* **23**, 561–563.
Heuse, G. A. (1957). "Biologie du Noir". Lielens, Brussels.
Hoagland, H. (1933). *J. Gen. Psychol.* **9**, 267–287.
Hirsh, I. J. (1959). *J. Acoust. Soc. Amer.* **31**, 759–767.
Hirsh, I. J. and Sherrick, C. E. (1961). *J. Exp. Psychol.* **62**, 423–432.
Kahneman, D. (1968). *Psychol. Bull.* **70**, 404–425.
Kibbler, G. O. and Richter, D. (1950). *Electroencephalog. Clin. Neurophysiol.* **2**, 227.
Kleitman, N. (1963). "Sleep and Wakefulness". Univ. of Chicago Press, Chicago.
Kristofferson, A. B. (1965). *NASA Contract. Rep.* No. 194.
Kristofferson, A. B. (1966). *NASA Contract. Rep.* No. 427.
Kristofferson, A. B. (1967). *Science* **158**, 1337–1339.
Kristofferson, A. B. (1969). McMaster University Technical Report No. 36.
Landauer, T. K. (1962). *Percept. Mot. Skills.* **15**, 646.
Lansing, R. W. (1957). *Electroencephalog. Clin. Neurophysiol.* **9**, 497–504.
Latour, P. L. (1967). *Acta Psychol.* **27**, 341–348.
Legg, C. F. (1968). *J. Exp. Psychol.* **78**, 46–49.
Le Grand, Y. (1937). *Acad. Sci. Paris C.R.* **204**, 1590.
Lichtenstein, M. (1961). *Percept. Mot. Skills* **15**, 646.
Lichtenstein, M., White, C. T., Siegfried, J. B. and Harter, M. R. (1963). *Percept. Mot. Skills.* **17**, 523–536.
McGill, W. J. and Gibbon, J. (1965). *J. Math. Psychol.* **2**, 1–18.
McReynolds, P. (1953). *Psychol. Rev.* **60**, 319–330.
Miller, G. A. (1951). "Language and Communication". McGraw–Hill, New York.
Mundy-Castle, A. C. (1962). *In* "Medical and Clinical Aspects of Ageing" (H. T. Blumenthal, ed.), pp. 575–595. Columbia University Press, New York.
Mundy-Castle, A. C. and Sugarman, L. (1960). *Electroencephalog. Clin. Neurophysiol.* **12**, 895–904.
Murphree, O. D. (1954). *J. Exp. Psychol.* **48**, 57–61.
Obrist, W. D. (1965). *In* "Behaviour, Ageing and the Nervous System" (A. T. Welford and J. E. Birren, eds). Charles C. Thomas, Springfield, Illinois.
Pitts, W. and McCulloch, W. S. (1947). *Bull. Math. Biophys.* **9**, 127–147.
Rémand A. and Lesèvre, N. (1957). *Electroencephalogr. Clin. Neurophysiol.* Suppl. **6**, 235–255.
Roufs, J. A. J. (1963). *Vision Res.* **3**, 81–91.
Saltzman, I. J. and Garner, W. R. (1948). *J. Psychol.* **25**, 227–241.
Seashore, C. E. (1938). "Psychology of Music". McGraw-Hill, New York.
Shallice, T. (1967). *Brit. J. math. Statistical. Psychol.* **20**, 129–162.
Sternberg, S. (1969). *Acta Psychol.* **30**, 276–314.
Stroud, J. M. (1955). *In* "Information Theory in Psychology" (H. Quaster, ed.), pp. 174–207. Free Press, Glencoe, Illinois.
Surwillo, W. W. (1961). *Nature (London)* **191**, 823–824.
Surwillo, W. W. (1963). *Electroencephalog. Clin. Neurophysiol.* **15**, 105–114.
Surwillo, W. W. (1964a). *Electroencephalog. Clin. Neurophysiol.* **16**, 510–514.
Surwillo, W. W. (1964b). *Electroencephalog. Clin. Neurophysiol.* **17**, 194–198.

Thor, D. H. and Spitz, H. H. (1968). *Psychon. Sci.* **13,** 291–292.
Vaughan, H. G. Jr., Costa, L. D. and Gilden, L. (1966). *Vision Res.* **6,** 645–656.
Venables, P. H. (1960). *Brit. J. Psychol.* **51,** 1.
Walsh, E. G. (1952). *J. Physiol. (London)* **118,** 500–508.
Walter, W. G. (1950). *J. Ment. Sci.* **96,** 1–13.
Welford, A. T. (1958). "Ageing and Human Skill". University Press, Oxford.
Werboff, J. (1962). *Exp. Neurol.* **6,** 152–160.
White, C. T. (1963). *Psychol. Monogr.* **77,** No. 12. (Whole no. 575).
White, C. T., Cheatham, P. G. and Armington, J. C. (1953). *J. Exp. Psychol.* **46,** 283–287.
White, C. T. and Harter, M. R. (1969). *Acta Psychol.* **30,** 368–377.
Wiener, N. (1948). "Cybernetics". Wiley, New York.
Zubek, J. P. and Welch, G. (1963). *Science* **139,** 1209–1210.

CHAPTER 6

# Menstrual Cycles

JUNE A. REDGROVE

*Department of Part-Time Business Studies and Management, Enfield College of
Technology, Enfield, Middlesex, England*

## I. Introduction

All females of childbearing age are subject to regular physiological
changes associated with the reproductive cycle. In humans, the cycle
begins with the development of an ovum in the ovaries and continues with
its release. If pregnancy does not ensue, the cycle ends with menstruation.
Thus, the cycle has 2 main points, ovulation and menstruation. For
convenience it is often divided into 4 phases, pre-ovulatory, ovulatory,
pre-menstrual and menstrual. A detailed account of the physiological
changes which take place during each of these phases is given in Appendix
I.

From the dawn of civilization taboos and superstitions have surrounded
the subject of menstruation. Taboos regarding menstruating women were
prevalent in primitive societies and ranged from forbidding women to
touch tools and utensils to complete banishment from the village for the
duration of the menstrual period. It is easy to understand this when one
considers that not only is the menstrual period somewhat dramatic, in
that it is accompanied by considerable bleeding, but also its periodicity
appears to correspond to phases of the moon. A delightful legend from

New Guinea claims that in ancient times the moon lived on earth in the form of a handsome young man, who made repeated attempts to seduce innocent young maidens. Eventually a young wife succumbed to his entreaties, but, unfortunately, her husband discovered them and set fire to the house in anger. The young man was trapped and killed; his blood spurted to the heavens in a great stream and there it became the moon. His revenge was to cause young women to suffer a loss of blood whenever he, the moon, appeared. Old and pregnant women were absolved from this fate, because he accepted responsibility for their condition (Fluhmann, 1956). Traces of this legend can be detected in our own society, in that "the curse" is a prevalent, popular name for the menstrual period and several studies have investigated the possibility of some relationship between menstruation and phases of the moon (Bramson, 1929; Gunn, 1937; Menaker, 1959).

The taboos on menstruating women derive to a large extent from the belief that menstruation exists for the purpose of removing poisonous substances from the body and that these substances will ruin anything touched by women during menstruation. In France, at one time, menstruating women were banned from sugar refineries, because it was believed that their presence would cause sugar to turn black. In 1878, in the *British Medical Journal*, a physician reported 2 instances where hams had been spoiled because menstruating women had cured them. There is some scientific support for these beliefs. MacKinnon (1954) found changes in palmar sweat activity associated with the menstrual cycle, and Fluhmann (1956) records the case of Professor Schiff, whose secretary placed a fresh rose on his desk every morning. He noticed that on some days each month the rose faded very rapidly and he noted the dates. When he checked these with his secretary he found that they corresponded with the dates of her menstrual periods.

Not surprisingly, there is a common belief (which persists to the present day in most societies) that women are significantly affected by the menstrual cycle, both physically and mentally. Indeed, periodic illness, absence from work and irritability are considered by many to be unavoidable events of nature. The purpose of this chapter is to examine the evidence for changes, and to see whether they have any real effects on behaviour in general and on performance in particular. It will also consider whether these effects can be controlled.

## A. *The menstrual cycle and behaviour*

1. *Psychological changes.* Apart from those changes which go to make up the menstrual cycle, there is considerable evidence of changes in bodily activity. Blood pressure (Eagelson, 1927; Moore and Cooper, 1923;

Seward, 1934), metabolic rate (Benedict and Finn, 1928; Gorkine and Brandis, 1936; Govorukhina, 1964; Hitchcock and Wardwell, 1929; Rubenstein, 1938; Seward, 1934), pulse rate (Eagelson, 1927; Kleitman, 1949; MacKinnon, 1954; Seward, 1934), body temperature (Abramson and Torghele, 1961; Fluhmann, 1956; Harvey and Crockett, 1932; MacKinnon, 1954; Marshall, 1963; Morton, 1950; Rubenstein, 1938; Seward, 1934) and body weight (Abramson and Torghele, 1961; Bruce and Russell, 1962; Morton, 1950; Tinklepaugh and Mitchell, 1939) have all been found to vary in relation to the menstrual cycle. Although there is considerable disagreement with regard to the direction and extent of these relationships, in general, bodily activity, as indicated by the above factors, rises during the pre-menstruum and falls to subnormal levels with the onset of menstruation.

In studies of the relationships between the menstrual cycle and psychological phenomena, explanations are frequently found in terms of sexual feelings and orientation. There are probably 2 major reasons for this. Firstly, there is a natural tendency, when linking behaviour and a sex cycle, to think in terms of sexual behaviour. Secondly, several of the studies were carried out by psychoanalysts, who tend to describe behaviour in these terms. However, they have made important contributions to work in this field.

In one of the first studies, Benedek and Rubenstein (1939a, 1939b) attempted to relate psychodynamic processes to ovarian activity. Nine subjects (hospital patients) were studied daily, for between 4 and 15 cycles. Seventy-five cycles were studied in all. Benedek, a psychoanalyst, compiled psychoanalytic records and Rubenstein, a physiologist, analysed vaginal smears. Each independently assessed the stages of the menstrual cycle the subjects were at. When they met for the first time at the end of the investigation they found complete agreement in their estimates. The significance of this result was that it demonstrated that phases of the menstrual cycle could be identified on the basis of the analysis of psychological variables, which indicates a clear relationship between behaviour and the menstrual cycle.

Altmann, Knowles and Bull (1941) studied the psychosomatic condition of 10 mature college women daily, during 55 menstrual cycles. Psychological factors such as mood, tension and irritability were estimated during interviews with subjects. Two phases marked by outbursts of physical and mental activity were observed during the cycle. One was the premenstrual phase, which was accompanied by irritability and tension; the other occurred around the time of ovulation, and in this case these symptoms were markedly absent. Benedek and Rubenstein (1939a) observed "a relaxed feeling of well-being" during ovulation; this agrees with the findings of McCance, Luff and Widdowson (1937), who studied emotional periodicity in women by asking them to fill in daily question-naires. In general, it would appear that euphoria and elation are con-comitants of ovulation, although increased tension (Greene and Dalton, 1953) and psychosomatic symptons (Abramson and Torghele, 1961) have been observed at this time.

2. *Pre-menstrual tension (PMT)*. Tension in the pre-menstrual phase of the cycle has been observed with sufficient frequency and severity to result in the term pre-menstrual tension (PMT) being recognized as describing the syndrome of symptoms which are to be found in many women during this phase of the cycle. "The point where they emerge from simple physiologic responses to pathologic processes is beyond definition" (Fluhmann, 1956). Differences in definition give rise to variations in the estimated incidence of PMT. Fluhmann (1956) suggested that 60% of all normally menstruating women experience mild or severe symptoms. Lamb (1953) observed symptoms in 73% of 127 student nurses. In a study involving 61 normals and 84 psychotics Rees (1953) found symptoms of PMT in 40·4%; in 15·6% these were severe, and the incidence among psychotics was greater than among normals.

In studies which included only those cases severe enough to warrant medical attention, 27% of the inmates of a women's prison (Dalton, 1961) and 26% of a sample of 715 normal women (Dalton, 1954) were found to be suffering from PMT.

These estimates indicate that approximately 1 woman in 4 is severely handicapped at some point in the menstrual cycle and that approximately 2 out of 3 experience some disturbances.

The pre-menstrual syndrome has been fully reviewed by Dalton (1964). In an early paper, Israel (1938) described it as follows: "It occurs in women between the ages of 20 and 40 years and is characterized by a cyclic alteration of personality. This alteration appears abruptly from 10 to 14 days prior to the expected menstruation and terminates dramatically with the onset of flow. The monotonous periodicity of the syndrome and its precursive relationship to the menses are striking phenomena. The illness regularly begins as a dire foreboding sensation of indescribable tension. The patient often inadequately describes this sensation by saying that she 'would like to jump out of her skin', and her feeling is manifest by her unusual behaviour. When the tension periodically reaches its maximum height, the manic activity of the patient beggars description. There are marked physical unrest and constant irritability . . . ."

PMT is thought to be caused by increased water retention which results in oedema. This gives rise to feelings of bloating and tightness of the skin, which lead to irritability and tension (Greenhill and Freed, 1940). Increased water retention has been variously attributed to an oestrogen-progesterone imbalance (Frank, 1931; Freed, 1945; Gilman, 1942; Kroger and Freed, 1951; Morton, 1950; Reich, 1962), pitressin or ACTH disturbance (Bickers, 1951) and numerous other factors (Bruce and Russell, 1962).

It has been observed that anti-social activity occurs more frequently in the menstruum and pre-menstruum than at other phases of the menstrual cycle. Several investigators found that suicide occurred most frequently during menstruation (Balasz, 1936; Rodrigues da Costa Doria, 1947; Rosenzweig, 1943), particularly on the first day of flow. The incidence of crime increased both menstrually and pre-menstrually (Cooke, 1945; Dalton, 1961; Morton, 1953; Rodrigues da Costa Doria, 1947), and schoolgirls tended to misbehave more during menstruation (Dalton,1960a). Individual cases of kleptomania during menstruation (Middleton, 1934) and the pre-menstruum (Lederer, 1963) have also been reported. In a study of 100 young women Schwarz (1959) found poorer overall adjustment, greater emotional lability and egocentricity, loss of consideration for others and a decrease in the capacity for planning, organization and integration, during the menstrual period.

## B. *The menstrual cycle and performance*

The evidence of psychological changes during the menstrual cycle supports the view that there is a tendency for women to feel elated at the time of ovulation and depressed during the pre-menstruum or at the start of the menstrual period. It is also clear that some women, who suffer from pre-menstrual tension, experience relief at the onset of menstruation. If feelings of euphoria and elation increase the capacity to perform jobs and depression reduces it, then on the basis of the evidence one would expect lower performances during the pre-menstrual and menstrual periods, with higher performances during menstruation in some women who suffer from pre-menstrual tension. There should also be higher performances during ovulation.

Broadly, the evidence concerning changes in performance during the menstrual cycle follows these patterns. In general the results fall into 3 groups, which give lowest performances during (1) menstruation, (2) the pre-menstruum and (3) at regular intervals throughout the cycle.

1. *Menstruation.* Lewin and Freund (1930) tested 12 subjects in order to assess the speed and quality of work and the ability to persist at tedious tasks. During menstruation the quality of work remained constant and there was a tendency for speed to increase, although persistence decreased. The authors' interpretation of these results was that there was a decrease

H

in skill, which had been offset by increased effort. Johnson (1932) observed 34 women learning to walk a tight-wire, 3 times a week. There was a precipitous drop in the learning curve at the onset of menstruation, followed by a peak at about the eleventh day of the cycle. He concluded that during the menstrual period there was greater difficulty in co-ordinating the relevant muscle groups.

Gorkine and Brandis (1936) found decreases in industrial performance during menstruation. They reported lower outputs (which they ascribed to more periods of inactivity) among 80 women doing light assembly work in the electrical industry. In a survey of 7867 women, diminished output was found among those remaining at work on the first day of the menstrual period (Anon, 1930). Dalton (1968) studied G.C.E. examination performance among 159 schoolgirls. Lower pass rates, distinction rates and average marks were obtained when examinations were taken during menstruation or the pre-menstruum. The decreases were most significant for girls with long cycles or long menstrual periods. Vernon and Parry (1949) studied performance on selection tests among women in the British forces. No relationships between menstruation and performance were observed for mental tests, but on practical tests there was a tendency for performance to improve at this time. These results were supported by Wickham (1958), who carried out similar studies. They suggest that various aspects of performance may be affected differently by the menstrual cycle.

2. *The pre-menstruum.* Studies which have found evidence of lowered performance during the pre-menstruum have also noted rises at the onset of menstruation with maxima in the intermenstruum (the phase between the end of the menstrual and the beginning of the pre-menstrual periods). Eagelson (1927) observed such changes in steadiness, hand-eye co-ordination and ability in simple mathematical tasks, in 3 women. Lough (1937) administered learning tests on 30 consecutive school days to 65, and on 40 consecutive school days to 31, student teachers. Results were analysed after having been grouped according (a) to the 4 phases of the menstrual cycle, and (b) to certain specific days, namely the first and second days of menstruation, the first day of the post-menstrual period and the mid-point of the cycle. No difference in progress was observed at different phases of the cycle, and mental activities requiring only speed were unaffected by menstruation. Accuracy increased significantly during the second day of the menstrual period. Herren (1933) measured the 2-point threshold for pain, touch and tactile sensitivity in 5 women for a total of 11 cycles. Recordings were taken 5 days prior to the onset of menstruation, 3 days following cessation and on a day 2 weeks after the

onset of the previous menstrual period; pre-menstrual readings were lower than the other two, but no indication of significance levels was given.

Evidence of lower industrial performance during the pre-menstruum has been reported by Sfogliano (1964). He found lower outputs among women making transistors in 59 out of 130 cases (43·3%). In 3 of these output dropped by 50%, in 29 by 40%, in 8 by 20% and in 15 by 10%. Morton (1953) surveyed women working in a prison laundry for symptoms of PMT. When these women were treated by him for their symptoms output increased by approximately one third.

In 2 independent studies of sporting performance, the pre-menstruum appeared to be a phase of lowered efficiency. In one study, involving 30 women between the age of 17 and 22, performance fell pre-menstrually and rose during the menstrual period (Brunelli and Rottini, 1965b). In the other study, performance was lowest during the pre-menstruum in 50 young female atheletes (Fichera and Romano, 1965).

3. *Changes throughout the cycle.* Sowton and Myers (1928) tested 29 women daily, on a series of mental and motor tasks, for approximately 6 menstrual cycles. The results were dealt with in 2 different ways. In one, "compressed curves", the cycle was divided into units of length equal to the length of the menstrual period; mean performances were calculated for each unit. In the other, "composite curves", cycles were adjusted to an arbitrary length (30 days) by adding or subtracting days in the intermenstruum; mean performances for each day of the 30-day cycle were calculated. Compressed curves gave mean performances for large units of the cycle (equal to the length of the menstrual period), thus providing a basis for comparing performance during the menstrual period with that during the rest of the cycle. Composite curves provided mean daily performance scores so that variations in efficiency could be observed throughout the cycle. As much variation was observed during the menstrual period as during the rest of the cycle, although periodic variations were observed when the results were analysed by composite curves. Farris (1956) compared daily performance with the menstrual cycle in 2 groups of women. Group I comprised 7 women who worked in a shell-loading factory. The job consisted of weighing powder, placing it in shells and stacking the packed shells in trays of 24; the operators were paid on a piece work basis. Group II consisted of 3 women stripping spools of film in a photographic dark-room; wages depended on the number of spools stripped per day. The operators recorded the first day of menstruation for at least 2 consecutive cycles. Only the results of Group I were reported in detail, because both gave similar results. These indicated 3 peaks in the level of output,

occurring on the fourth, twelfth and twenty-fifth days of a cycle of mean length 30 days.

These studies suggest that there may be significant daily changes in performance throughout the menstrual cycle, which may be hidden if the cycle is dealt with in too large units. Further support is given to this hypothesis by the results of Smith (1950a, 1950b), who studied 86 women in 3 factories for between 2 and 3 cycles. Some significant changes in efficiency were observed when comparisons were made between 4 phases of the cycle, but not when they were made between menstrual and non-menstrual days.

4. *Individual differences*. From the evidence so far it would seem that there are individual differences in the effects of the menstraul cycle. Some women feel worst during the pre-menstruum, whilst others experience difficulties during menstruation. Sowton and Myers (1928) observed marked individual differences during the course of their investigations, and Kirihara (1932), in an industrial study involving 110 women doing several different jobs, found that individual differences prevented generalizations as to the exact time of the occurrence of decreases in performance levels which corresponded to the menstraul cycle.

Pierson and Lockhart (1963) were unable to find significant changes in reaction time in a study of 25 women. Each subject was tested 4 times, 2 days before menstruation and 2, 8 and 18 days after, but, in the analysis a **mean** reaction time from all subjects was calculated for each day sampled. In this study it is possible that the individual differences in the timing of changes cancelled each other out so that no significant relationships could be observed.

## II. Some Experiments

The evidence concerning relationships between performance and the menstrual cycle varies and is sometimes contradictory. Eayrs and Glass (1962) suggested that this might be due to variations in techniques of investigation and analysis. Several methodological issues need to be considered in this context. Of fundamental importance are questions such as the extent and nature of individual differences, the units in which the cycle should be considered (day or phase), methods for treating cycles of different length so that they are comparable, and the accuracy of various reference points and methods for determining them. Some experiments were carried out in an attempt to answer some of these questions. In order to do this it was necessary to obtain sufficient data for individuals, on a daily basis, to permit detailed analysis. In all, 3 experiments were done: a pilot study; one involving a typing operation, where performance was based on specific tests set by the investigator; and one of a punch-card operation, where daily work performances of the subjects (as recorded by the firm for calculating wages) provided the data (Redgrove, 1968).

## A. *The pilot study*

**1. *The operations*.** The pilot study involved 8 women working in a laundry and covered a period of 5 months. Two operations were studied: sorting and marking; and pressing white coats.

Sorting and marking was a complex operation demanding both mental and motor skill. All articles entering the laundry had to be checked, sorted and marked before they were washed. Each article was dealt with individually. To carry out the task, the operator picked up a bundle of work, unwrapped it and put the customer's book on a stand in front of her. Working through the customer's list, she marked the articles, one at a time, and then shook them into a bin from which they were removed by another operator for washing. The operator recorded the number of bundles and individual items marked on her work sheet.

White coats were pressed on a 3 press unit. The operation was designed so that, by placing parts of coats on the presses in a predetermined order, 6 coats were pressed simultaneoulsy. The presses were synchronized so that as press 1 closed, press 2 opened and so on. The operator worked at each press in turn.

Wages were paid on a standard unit time basis, i.e., piece work. Where performances were below standard or there was waiting time operators were paid at a standard rate; thus on slack days performances would be weighted towards standard. The amount of work available was subject to both long and short term variations. Most work was based on a three-day contract so that schedules ran as follows:

collect Monday     —return Wednesday
collect Tuesday    —return Thursday
collect Wednesday—return Friday
collect Thursday   —return Saturday.

This meant that work built up at the beginning of the week in departments at the beginning of the laundering process and towards the end of the week in those concerned with the later stages. Sorting and marking was at the beginning and pressing white coats was near the end of the operational sequence.

Long term fluctuations were not as specific as weekly ones. The laundry trade is to some extent affected by the weather and so the demand for service tends to be as unpredictable as the weather.

2. *Collection of data.* Daily work performance figures for the operators participating in the study were obtained directly from the firm's records. Menstrual data were obtained from questionnaires, which the operators filled out every working day. Sowton and Myers (1928) suggested that knowledge of the purpose of the study might bias the results and recommended that subjects be unaware of the topic under investigation. Following this recommendation, the study was disguised as an investigation of the health and attitudes of women at work and their influence on performance.

In the first instance the operators were approached as a group and, later, individually by letter and interview. The group and letter approach gave the operators some idea of what was to be required of them, so that when they came to be interviewed they were aware of the circumstances and any difficulties or misunderstanding could be dealt with. This approach helped to build up the operators' confidence in the investigator and made it possible to meet any opposition early in the study. Only one operator refused to co-operate, and she was not pressed to do so, since it was thought that this might induce resentment on her part which could be transmitted to her colleagues.

The principles governing the design of the questionnaire were simplicity and brevity. Although the operators were of average intelligence, it was undesirable that any ambiguities or difficulties in interpretation be introduced. The supervisor agreed to introduce the filling up of forms into the normal daily routine, but they were not in fact completed until the end of the day. The answers were, therefore, possibly affected by the day's happenings.

Pre-dated forms were taken to the laundry once a fortnight. On these regular visits subjects were encouraged to continue filling out the forms, and any problems arising from previous replies were sorted out. This meant that points were clarified very soon after they were recorded. This was particularly important with regard to dates of onset of menstruation, because, where menstruation was first recorded on a Monday, the menstrual period could have started Friday night, Saturday, Sunday, or during the work period on Monday. The investigator had to check with the operators the actual date on which the periods started, so that days of the menstrual cycle could be accurately demarcated.

3. *Results and discussion.* Data were collected for 25 complete cycles from 6 women sorting and marking (including the supervisor), and 2 women pressing white coats (see Table I). S4 reached the menopause and S5 became pregnant 2 months after the start of the investigation. One of the girls pressing white coats (P2) recorded only 1 menstrual period, although she completed the questionnaires regularly.

She claimed that she was menstruating regularly, but her periods apparently started over week-ends. When she was asked the actual dates, she said she couldn't remember them. She was a very self-conscious girl and it is possible that she felt reluctant to disclose the dates of her menstrual periods.

TABLE I

*Age, average cycle length and the number of cycles completed by each subject*

| Operator | Age (years) | Average cycle length (days) | Number of complete cycles |
|----------|-------------|------------------------------|----------------------------|
| S1 | 29 | 26·2 | 5 |
| S2 | 37 | 29·0 | 5 |
| S3 | 40 | 24·8 | 4 |
| S4 | 45 | 25·7 | 3 |
| S5 | 28 | 25·5 | 2 |
| S6 | 44 | — | — |
| P1 | 22 | 27·2 | 5 |
| P2 | 16 | 29·0 | 1 |

(a) Performance Scores. At the beginning of the study it was thought that a time series type of analysis would show any periodic trends which correspond with the menstrual cycle. One advantage of this would be to test for regular changes in post-menopausal women (Torghele, (1957) found evidence of periodic emotional disturbance, comparable to PMT, in post-menopausal women, and S6 was included in the sample to investigate this further). However, because the phases of the menstrual cycles did not occur on a regular time basis a time series type of analysis was not possible. For this reason, and because of the lack of adequate menstrual data from subjects S4, S5 and P2, only the performance data of subjects S1, S2, S3 and P1 were subjected to the further analysis described below.

Because of the weekly trends in the supply of work, the relationship between performance and day of the week was first examined by separate analyses of variance of each subject's scores. In all 4 cases the relationship was significant (in 7 cases $p < 0.01$ and in the fourth $p < 0.05$). The weekly trend was therefore removed by subtracting mean performances for the appropriate days of the week from raw scores, and the significance of the relationship between days of the menstrual cycle and these adjusted performance scores was then examined by further analyses of variance. No significant relationship was found in any of the 4 cases.

(b) Questionnaire Responses. Patterns of response to the questionnaire varied considerably among the 8 subjects. Some tended to answer "yes" to the question "Do you feel happier than usual?" nearly every day. Others made very few positive responses. In analysing the nature of

responses in relation to the menstrual cycle, the number of items responded
to appeared to be the most profitable approach, on the basis that overall
responsiveness to the questionnaire, rather than to specific items, was
more indicative of changes in feelings. Days on which more than 1 (or 2,
depending on the general frequency for each subject) positive response
was made to items on the questionnaire were termed "critical events".
The occurrence of these events in relation to the menstrual cycle was
examined for subjects S1, S2, S3, S4 and P1 by the Chi-square test.

In 1 subject (S3) there was a significant relationship ($p<0.01$); of the
remaining 4, 2 were not significant and 2 recorded insufficient critical
events for analysis.

The results of this pilot study suggest that there was no relationship
between the menstrual cycle and **performance**, although in 1 case there
was a significant relationship between the cycle and **response to the
questionnaire**. Taking these findings together with those of previous
studies it seems likely that the present result arose either because the data
were insufficient or inappropriate, or because the effects of the menstrual
cycle were overshadowed by other, unknown factors in the work situation.
It is impossible to say which of these explanations applies. However, the
need has been demonstrated for long term studies which provide sufficient
data for each subject, as free as possible from the influence of extraneous
variables.

The study also revealed some points with regard to the collection of
menstrual data. Sowton and Myers (1928) suggested that subjects should
be unaware of the purpose of the study, in order to avoid bias in their
performances. If the effects of menstruation are being studied, scores from
days on which menstruation is recorded can be compared with those
obtained on non-menstrual days, but where the whole cycle is being
investigated it is necessary to locate cycle days as accurately as possible.
It is difficult to achieve sufficient accuracy without subjects becoming
aware of the purpose of the investigation. In the pilot study dates of onset
of menstruation were obtained indirectly from the questionnaires. Where
menstruation began during a consecutive series of daily recordings the
first day on which menstruation was reported was taken as day 1 of the
cycle. However, on several occasions flow began during a gap in records
and it was not clear when menstruation had actually started. In these
cases the operators were asked for the actual dates of onset, and it is probable that because of these enquiries the operators attached considerable
importance to menstruation in the context of the investigation, so that
the purpose of the enquiry was not effectively concealed.

If performance is to be analysed by day of the menstrual cycle, the latter
must be determined accurately in terms of data and time of onset of men-

struation. Considerable errors can arise if only the date is recorded. For example, consider two cycles, one starting at 06.00 on October 3rd and ending at 22.00 on November 3rd, the other starting at 22.00 on October 3rd and ending at 06.00 on November 3rd. If only dates of onset are recorded, both cycles will be considered as 31 days long and October 3rd will be taken as the first day of the cycle. However, in the second case October 3rd is a pre-menstrual day up to 22.00, and with regard to work performance October **4th** should be taken as day 1 of the cycle. It would appear from this that it is difficult to get sufficiently accurate menstrual data and to keep subjects unaware of the purpose of the study. If a daily analysis is to be undertaken it is recommended that subjects be informed of the purpose of the study so that detailed information on menstruation can be obtained. Furthermore, if subjects do know the purpose of the investigation any bias is only likely to arise on menstrual days, because these are the only parts of the cycle of which the subjects are clearly aware. If relationships can be shown to exist between the **whole cycle** and performance, the importance of subjects knowing the purpose of the investigation is clearly unimportant.

## B. *The typing and punch-card studies*

The main purpose of these investigations was to examine daily changes in the performances of individual women during the course of several menstrual cycles. A sample size of 10 recordings per day of cycle was aimed at for each subject, which, allowing for week-ends, necessitated daily recordings for each subject for at least 1 year.

A major problem was to find a suitable task and situation. Laboratory tests were ruled out because it was impossible to find women willing and able to attend daily for such a long period without remuneration. However, women do go to their places of work daily (excepting week-ends) over long periods of time, and it was therefore decided to test them there.

It was also necessary to find a situation in which extraneous variables were kept to a minimum. This was somewhat difficult to achieve in practice, because of frequent variations in the product at many work places, and also because of seasonal factors which affect most jobs.

Eventually, 2 studies were undertaken. One concerned 3 secretaries, who were able to carry out special tests (typing set pieces each day) for at least 12 months. The other used daily performance scores of punch-card operators, which were collected as part of the normal work routine over a period of 10 months.

1. *Measurement of performance.* When typing tests are used as the basis for measuring performance they must be constant with regard to the degree

of difficulty, familiarity with the text, ratio of upper case to lower case letters and the extent to which various keys and fingers are used. It is essential to use different test pieces in order to avoid problems of learning and boredom which would probably arise if the same test was done day after day. Accuracy tests for the Royal Society of Arts' typing examinations were used as test material.

Unfortunately there were insufficient tests to cover the whole investigation without repetition. Twenty-five different tests were used, and each one was divided into 2, making 50 tests in all. The order in which tests were done was carefully controlled, so that repeats occurred after 50 working days (i.e. approximately 10 weeks) and it was unlikely that there was any significant recall of particular tests. Subjects were given a week's supply of tests at a time, which meant that they were seen regularly and could be encouraged to keep up with their records and tests. They were instructed to type as quickly and accurately as possible, without correcting errors. They timed themselves, so that they could fit tests easily into their work routines at their own convenience. They were asked to do the tests at the same time each day, as far as possible, in order to avoid changes in performance due to the effects of the time of day (see Chapter 2).

Speed and accuracy of typing were calculated as follows:

(a) speed $= \dfrac{\text{number of strokes}}{\text{time (sec)}}$

(b) accuracy $=$ number of errors.

The number of strokes was the number of key depressions involved in typing the tests, including space bar and carriage lift operations. There were difficulties in calculating errors, because at times the source of the error was not clear. For example, if "in" was typed instead of "on", it could have been due to a wrong key, or to misreading of the script; thus there was inevitably an arbitrary element in placing numerical values on errors.

The punch-card investigation took place in the check office of a Co-operative Society. Sales in all shops were recorded on small receipt vouchers, which were arranged in books of 50 pages, 20 vouchers to a page. The week number, branch number, member's share number and the amount received were recorded on each receipt. The check office collated this information, so that the total expenditure of each customer and the receipts of each branch could be calculated. In the operation studied, information was transferred from each voucher on to a punch-card. All the vouchers in a book had the same week and branch numbers, which were set into the machine and punched automatically, so that the operators punched only the member's share number and the amount spent for each voucher. Operators supplied themselves with work, taking it from piles spread around the work area. Branch returns were sent in weekly, and in an average week there was just enough work to last until Friday. Monday morning was spent unpacking new work.

The operators were paid on a bonus system, based on the number of cards punched per hour per week and on the number of errors. They kept their own records of the number of cards punched each day. Daily figures for the number of cards punched, minutes worked and errors made, constituted the raw data in this investigation. There were errors in these daily figures, because accuracy was only required over the week as a whole. Operators recorded the number of cards punched a book at a time, books which were partially completed at the end of a day being recorded for the following day. As there were 1000 vouchers per book, these errors were considerable. However, they were reduced by putting all data on a 3-day basis, and calculating 3-day moving averages for the number of cards punched per minute and the number of errors per 1000 cards.

Although skilled subjects were used in these studies to avoid long-term learning effects, it was thought possible that some other long-term changes might nevertheless have taken place in performance. When speed and accuracy were plotted on time scales for each subject, a slight increase in speed was observed for two of the typists, and in the punch card study one operator (B3) had scores outside the subsequent range for the first month of the investigation; these scores were therefore omitted. In 2 others (B5 and B6) performance varied within 2 or 3 distinct ranges over the period of the investigation. These long term trends were removed by adjusting typing scores to the mean for each cycle, and scores for the punch card operators to the mean of the particular range-group into which the data fell.

2. *Collection of menstrual data.* In the typing study subjects filled out record cards daily. Each card covered a calendar month, with spaces marked for each day. Each space was divided into sections headed: period, headache, stomach-ache, illness, medicine, depressed, happy, anxious, irritable, restless. Subjects were asked to tick the section if the heading was appropriate to their feelings on a particular day. There were also spaces for them to record any additional information which they thought might be useful. Ovulation dates were estimated from basal body temperature charts (Van de Velde, 1939). Subjects took their oral temperatures with a clinical thermometer every morning, immediately on waking. Tempera-

TABLE II

*Ages and menstrual data for subjects in the typing and punch-card studies*

| Operator | Age | Average cycle length (days) | Number of completed cycles |
|----------|-----|-----------------------------|----------------------------|
| T1 | 20 | 29·9 | 16 |
| T2 | 20 | 28·6 | 18 |
| T3 | 28 | 26·0 | 12 |
| B1 | 19 | 29·9 | 9 |
| B2 | 19 | 24·0 | 9 |
| B3 | 18 | 32·6 | 6 |
| B4 | 19 | 29·2 | 4 |
| B5 | 16 | 31·0 | 8 |
| B6 | 16 | 28·4 | 7 |
| B7 | 17 | 29·2 | 7 |
| B8 | 19 | 31·9 | 7 |
| B9 | 18 | 29·0 | 9 |

tures were recorded on cards which covered a calendar month, with spaces marked for each day.

Initially, subjects in the punch-card study were given cards similar to those used in the typing study, but they failed to fill them in every day. They were probably to some extent overwhelmed by the amount of information asked for, and it was thus decided to ask them just to record the dates and times of onset of menstruation, which they did quite satisfactorily.

3. *Results.* The ages and menstrual data of all subjects are summarized in Table II, and the ovulatory data for the typists are given in Table III. As a group the secretaries were older than the punch-card operators and more menstrual cycles were covered in the study. There were considerable variations in cycle length between individuals, the range being 24·0 to 32·6 days, with a median of 29·9 days. Thus there was a tendency for subjects in this study to have cycles longer than 28 days, which is usually considered as the average for all women. It is interesting to note that cycle lengths in the pilot study tended to be shorter than 28 days and the subjects were generally older. These findings are similar to those of Gunn (1937), who observed decreases in cycle length with age.

No common trends existed with regard to ovulation, which variously occurred before (T2), after (T3) and around (T1) the middle of the cycle.

The relationships between performance and the menstrual cycle were examined by analyses of variance, by day of the cycle for all subjects and by phase for the typists. (Phases were determined by dividing cycles into 4 equal parts, the first being the menstrual phase, days 1–7 in a 28-day cycle.) No significant relationships were found between

TABLE III

*Pre- and post-ovulatory intervals for the typists*

| Operator | Pre-ovulatory interval | Post-ovulatory interval |
|----------|------------------------|-------------------------|
| T1 | 14·3 | 14·8 |
| T2 | 12·6 | 16·0 |
| T3 | 13·5 | 12·5 |

accuracy and the menstrual cycle; however, so few errors were made by both sets of subjects that there was in any case little chance of detecting any effects of the menstrual cycle with these scores. In no instance in the punch-card study was speed significantly related to cycle day, but in the typing study, in 2 out of the 3 cases a significant relationship between speed of typing and day of the menstrual cycle was observed (T1, $p < 0.025$; T2, $p < 0.01$). No significant relationships between phase of the menstrual cycle and performance were observed.

The occurrence of critical events in the typing study (as determined by the number of positive responses to the questionnaire) was analysed in relation to the menstrual cycle by Chi-square. Significant relationships were observed in both of the 2 cases where there was a sufficient number of events for analysis (T2, $p < 0.001$; T3, $p < 0.005 > 0.001$).

These results raise many points, which will be discussed against the background of other studies in the next section.

## III. Discussion

The evidence from the typing study strongly supports the view that the menstrual cycle brings about changes in behaviour and performance, but no evidence of these relationships was found in the other study. These results could have arisen either because the punch-card operators were unaffected by their menstrual cycles, or because, as a variable affecting performance, the menstrual cycle was not significant in this particular situation owing to the nature of the demands made on the subjects by the job. In order to determine the importance of the menstrual cycle as a factor in the work situation, it is necessary to consider it against the backgrounds of capacity and job demands.

### A. *Performance, capacity and effort*

Studies of the relationship between performance and the menstrual cycle are usually based on observations of performance at various stages of the cycle. Implicit in this approach is the assumption that effort is constant throughout, or that some function or functions within the individual (e.g. visual capacity) are used to their limits because of the nature of the job. This is so because, unless the subject is working to capacity, if capacity decreases performance can be maintained by increased effort. The extent to which increased effort is possible depends on the motivation of the subject on the one hand and the demands of the job on the other.

1. *Motivation.* One of the differences between the punch-card and typing studies was the motivation of the subjects. Typists T1 and T2 worked in the same office and competed with each other to see who could work fastest. This meant that they were working at a very high level of motivation. In the punch-card study the operators were probably working to produce an adequate day's work and no more. The atmosphere in the office was one of reasonable activity, but the operators did not appear to be working flat out. In these circumstances it is clear that if the typists' capacity was reduced by the menstrual cycle performance would be also, but the punch-card operators always had some spare capacity so that they could put in extra effort and hence maintain performance.

Evidence of changes in effort during the menstrual cycle comes from Lewin and Freund (1930), who observed an increase in the speed of working but a decrease in output during menstruation, in 12 women. They attributed this result to increased effort, which partially offset a drop in capacity. Gorkine and Brandis (1936) observed a lowering of output during menstruation due to an increase in the number of rest pauses. This suggests that the women had to use more effort to maintain working speed, which, in turn, increased the need for rest.

Sport, particularly competitive sport, provides another example of highly motivated subjects who can be assumed to be working at their limits. In an Olympic final, for example, if competitors do worse than their previous best performances it is because either their physical condition or their mental attitude (over-anxiety) is reducing their efficiency. In these circumstances the menstrual cycle could be very significant. This has been appreciated by various sporting authorities, and research has been aimed at finding methods for shifting cycles so that athletes compete at the time for performances to be optimum (Brunelli and Rottini, 1965a; Cseffalvay, 1966). A major problem in this area is that it is still not clear which are the best and worst phases of the cycle. Brunelli and Rottini (1965b) studied 30 women between the ages of 17 and 22 and observed falls in performance pre-menstrually, with rises during the menstrual period. Fichera and Romano (1965) also observed lowest performances during the pre-menstruum in a study of 50 young female athletes. Brunelli and Rottini (1965a) suggest that best performances are achieved in the period immediately following menstruation and recommend that, where events are scheduled for the pre-menstruum, the cycle should be brought forward in order to ensure that the athletes are competing at the peak of their ability.

2. *Job demands*. In some jobs operators are required to use certain of their faculties to full capacity; in others, they are not. The more difficult or demanding a job is, in terms of the extent to which particular faculties are exercised, the more likely it is to be affected by the menstrual cycle. Thus Smith (1950a, b) observed significant differences in output, related to the menstrual cycle, in jobs of high mental difficulty, but not in those of average and lower difficulty. Sfogliano (1964) studied women forming and soldering transistors. The components were so small and the operations so fine that work was carried out under microscopes. In this instance operators were using their visual mechanisms to the full. Lower performances occurred in the pre-menstruum, with a rise at the onset of menstruation.

Other evidence of changes in capacity is quoted by Seward (1934), who found 2 studies in which narrowing of the visual field was observed 2 or 3 days before menstruation, with a gradual return to normal about the seventh day after the onset of flow.

Studies of thresholds seek to determine the limits of capacity with regard to certain functions. Significant changes in threshold should, therefore, be recorded during the menstrual cycle if it affects capacity. Glanville and Kaplan (1965) measured sensitivity to the taste of quinine and 6-*n*-propylthiouracil (PROP) in one or more cycles of 19 women. Subjects were tested at the same time each day, 3 days a week, for 4 to 9 weeks. Days of the menstrual cycle were demarcated by counting forwards and backwards from menstruation. Sensitivities in 3 phases of the cycle were compared: pre-menstrual, menstrual and post-menstrual. They observed changes in sensitivity which corresponded to the menstrual cycle, but there were considerable individual differences between subjects. The general tendency was for subjects to reach maximum sensitivity after the onset of menstruation on days 1 to 5. Vierling and Rock (1967) measured olfactory sensitivity to exaltolide in 73 student nurses. Data from different subjects were grouped and significant peaks were observed, one 17 days and the other 8 days prior to the onset of menstruation.

Herren (1933) measured the 2-point threshold for pain, touch and tactile sensitivity in 5 subjects for a total of 11 cycles. Pre-menstrual thresholds were lower than those for the menstrual and inter-menstrual phases. In another study of cutaneous sensitivity, Kenshalo (1966) measured cool thresholds for 3 subjects. Thresholds were significantly higher between menstruation and ovulation than after ovulation. He attributed the changes to cutaneous vasodilation, which he found to be caused by progesterone.

## B. *The menstrual cycle as an experimental variable*

Certain aspects of the results of the above experiments, and of other studies, show that the menstrual cycle is by no means a straightforward experimental variable. There are several approaches to the problem, which depend on whether the cycle affects capacity on a daily or a phase basis, the appropriate techniques and points of reference for demarcating cycle days, and whether there are individual differences which prevent the grouping of data from different subjects. All of these considerations determine how many subjects should be studied, to what extent (daily, once a week) and for how long.

1. *Days and phases.* Although significant relationships were found between performance and days of the menstrual cycle in the typing study, no significant effects were observed with regard to phase. Figure 1 shows changes in performance during the menstrual cycle for the 3 typists. Best performances seem to occur between the −20th and +10th percentiles (days 22–3 of 28 day cycles); lowest scores are found between the −40th and −20th and the 15th and 35th percentiles (days 18–22 and 5–10 of 28 day cycles). Thus high performances tended to occur in the late premenstruum and early part of the menstrual periods, whilst low performances were recorded in the early pre-menstruum and late menstruum. In this study phases were demarcated by dividing the cycle into 4 equal parts corresponding to days 1–7, 8–14, 15–21 and 22–28 of a 28 day cycle.

It is worth noting that the phase of highest performance was that which one would have expected to be lowest and vice versa. A possible explanation is that subjects over-compensated when they felt low, thus producing an increase in output, but this is not consistent with the suggestion that they were working at full capacity. It may be that during the pre-menstruum there is increased tension, and that this leads to higher states of arousal which in turn improves efficiency, but this is merely conjecture. One other possibility is that phases of the cycle have different effects on different types of task, a view supported by the studies of Vernon and Parry (1949) and Wickham (1958), who found evidence of lower performances on mental tests and higher scores on motor tests during the menstrual period.

However, what is important in Fig. 1 is that the variations in performance cut across normal phases and that significant variations were observed with the day, but not with the phase of the menstrual cycle.

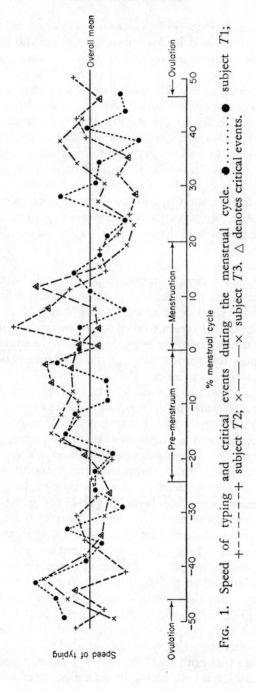

FIG. 1.  Speed of typing and critical events during the menstrual cycle. ●········● subject *T1*; +– – – –+ subject *T2*; ×– — —× subject *T3*. △ denotes critical events.

Again, Sowton and Myers (1928) compared menstrual and non-menstrual performances as well as daily variations throughout the cycle and found as much variation within the menstrual period as during the rest of the cycle, although there were periodic variations in performance. This suggests that perhaps the cycle was divided into phases on the wrong basis, although Smith (1950a, b) was able to show relationships between performance and phases of the menstrual cycle divided on this basis. Dalton (1964) has also obtained a large number of positive results, but her studies deal with different types of information and this will be discussed later. Thus although the method used divides the cycle into 4 phases, which correspond to the physiological changes (Appendix I), it is worth bearing in mind that another basis might be used.

Overall, the evidence supports daily changes in performance due to the menstrual cycle and it would seem that data should be collected on this basis.

2. *Demarcation of cycle days.* Contrary to popular belief the length of the menstrual cycle varies considerably. Extensive investigations have been unable to reveal a single case of perfect regularity (Arey, 1939; Gunn, 1937), and Holt (1935) likens a women with a perfectly regular menstrual cycle to a fairy-tale princess. An estimate based on the study of more than 30,000 cycles puts the normal range of the difference between two successive cycles at $\pm 6$ days (Seiichi, 1962).

Because of the variations in cycle length it is necessary to adjust cycles for comparison. This may be done on an absolute or proportional basis. To adjust cycles on an absolute basis, days are demarcated by counting $m/2r$ days backwards and forwards from reference points, where $m$ is the mean cycle length and $r$ equals the number of reference points available (usually only the dates of onset of menstruation are obtained, in which case $r = 1$). Cycles are adjusted proportionately by dividing them into a number of arbitrary units ($n$); Reynolds (1942) suggested 100. Days are demarcated by $ny/m$, where $y$ is the observed day of the cycle. The suitability of each method depends on the nature of the variation in cycle length and its causes. Where cycles stop and start at various points variation is absolute and it is proportional where cycles develop at different rates.

Studies of variations in cycle length have shown that there is more variation between cycles of different individuals than between cycles for one person (Arey, 1939; Gunn, 1937); that the first half of the cycle is more variable than the second (Marshall, 1963; Seiichi, 1962); and that severe trauma can arrest its development (Loeser, 1943). The implications of these findings are that both absolute and proportional types of variation apply in most cases, but that inter-individual differences are more proportional and intra-individual differences more absolute in nature and the appropriate method for demarcating cycle days should be selected accordingly.

One disadvantage of the proportional method is that it does not give data coincident on a time scale and extrapolation is necessary so that coincident data may be obtained for sophisticated statistical analyses. However, less elaborate tests can be made by using Chi-square. Here the occurrence of performances above and below certain limits at various stages of the cycle is compared with what would have arisen by chance.

A further difficulty in the demarcation of cycle days arises from the paucity of reference points. Ideally, each stage of the menstrual cycle would be determined by daily hormone assays (Benedek and Rubenstein, 1939). However, these techniques are tedious and very expensive and in most studies menstrual data consist of dates of onset of menstruation, from which days of the cycle are demarcated by the methods outlined above. Because of the variations in cycle length errors arise and thus it is desirable to get as many reference points as possible in order to reduce errors of demarcation.

Van de Velde (1939) outlined a method for determining dates of ovulation from basal body-temperature charts. He noticed a thermal shift (a clear rise) in temperatures around the time of ovulation, and in the typing study subjects completed basal body-temperature charts to provide some estimates of ovulation so that the feasibility and reliability of reference points thus determined could be investigated. Because the temperatures have to be basal (i.e. taken immediately on waking every morning) the subjects had to be able to take their own temperatures. This means that fairly intelligent subjects are required. The typists provided excellent charts from which ovulation dates were estimated. The reliability of these estimates as points of reference for demarcating data was examined by comparing the between-days and within-days variations in performance when mid-cycle data were demarcated separately in relation to the two reference points, onset of menstruation and ovulation. This was based on the assumption that within-days variation would vary in inverse proportion to the accuracy of the reference points. It is less likely that performances will be attributed to the wrong cycle days so there will be less within cycle days variation. Thus, when comparisons of the variance ratios are made, the more accurate reference points should yield higher variance ratios. Comparisons of the variance ratios using the two reference points are given in Table IV. In each instance variance ratios were greatest when menstruation was the reference point for demarcating mid-cycle data. For subjects T1 and T2 the level of significance was lower when ovulation estimates were used to demarcate mid-cycle data.

TABLE IV

*Variance ratios from 2 points of reference for mid-cycle data*

| | Point of reference | |
| Subject | Menstruation | Ovulation |
|---|---|---|
| T1 | 1·36 | 0·66 |
| T2 | 1·86 | 1·64 |
| T3 | 1·41 | 1·12 |

Marshall (1963) and Fluhmann (1956) have suggested that basal body temperature charts are inaccurate as indicators of ovulation, and, taking this together with the above results, It seems that this technique provides less accurate points of reference than dates of onset of menstruation.

3. *Individual differences.* It was suggested in section I.B.4 that individual differences in the effects of the menstrual cycle were likely to invalidate comparisons of data from different individuals. Sowton and Myers (1928) had observed considerable individual differences, but in the typing study, although there were some differences between individuals there did appear to be some common trends in performance during the cycle (Fig. 1). In Fig. 1 data were demarcated **proportionately** by transferring cycles on to a percentile basis; Sowton and Myers demarcated their cycles on an **absolute** basis. Thus the individual differences observed by Sowton and Myers may have resulted partly from the demarcation method they used, since subjects with cycles of different length would be compared, each day being taken in relation to its distance from the onset of menstruation. This may not give comparable bases in terms of cycle stages, because these occur at various distances from the date of onset of menstruation.

4. *Critical events.* The evidence above suggests that the menstrual cycle affects capacity on a daily basis and that there are individual differences in the timing of changes, which are mainly brought about by differences in cycle length. Thus similar stages of the cycle affect women in like manner, but these stages arise at varying distances from the date of onset of menstruation, depending on the average cycle length for each subject. The implications for experimental design are that different women should not be grouped together (particularly if cycles are demarcated on an absolute basis), nor should phases be the base unit into which cycles are divided.

Dalton (1964) has carried out a large number of studies in which she has examined the incidence of various phenomena in relation to the menstrual cycle, and she has obtained significant results by recording behaviour of an all-or-none type (critical events), for example, crime, suicide and accidents. A major difference between such critical events and performance as experimental variables is that the former refer only to positive instances, whilst in the latter case high and low values tend to cancel each other out. Thus a woman of 24 who committed suicide during the pre-menstrual period, and who had menstruated regularly from the age of 14, would have experienced approximately 120 pre-menstrual periods in which she did not commit suicide; these would not count. If, however, one were to deal with performance on the same basis, even if she had 30 very bad performances in the pre-menstrual period there would still be 90 average ones and there would therefore be a tendency for a mean close to the overall mean to be recorded.

Thus, for this type of analysis, Dalton's technique is satisfactory and as far as grouping individuals is concerned is the only possible method. For even the most anti-social woman gets relatively few opportunities to commit sufficient crimes in her lifetime to support analysis on an individual basis. With regard to suicide the point can be even more forcibly demonstrated. In this context it is important to note that the investigator has no control over sample size, other than by continuing the investigation until sufficient events have been recorded. Persuading subjects to commit crimes (even in the interests of scientific research) is not an acceptable way of collecting data, apart from problems of validity in these circumstances.

## IV. Conclusions

It seems that the menstrual cycle does affect the capacity to carry out certain tasks, but that the extent to which these effects will be manifest in changes in performance depends on the extent to which the decreases in capacity can be offset by increased effort. This has implications both for research and for problems arising because of changes in output in industry.

With regard to research aimed at determining the effects of the menstrual cycle on capacity and performance, it is necessary to study situations where the one reflects the other as closely as possible. This, it has been demonstrated, occurs where subjects are working at limits, either because they are highly motivated, or because certain faculties are used to the full. The study of thresholds is a very useful instance of the latter. Such research is necessary so that the effects of the menstrual cycle can be fully known and understood and so that methods for dealing with these effects can be developed where necessary.

With regard to overall work performance, the implications are that controlling the menstrual cycle is only likely to result in significant improvement in circumstances where the subjects normally work close to the limits of their capacity. In practice this is where particular faculties are extensively used, that is, where jobs present considerable visual, auditory, tactile and other similar demands. However, benefits may also be gained from considering the menstrual cycle in relation to situations which are particularly sensitive with regard to the possibility of accident. Dalton (1960a) found significant increases in the incidence of accidents during the pre-menstruum and menstruation. If individuals are at increased risk for, say 3 days per menstrual cycle, then there is a 1 in 9 chance that a critical situation will arise on a high-risk day. This, taken together with the likely cost of accidents in terms of money and danger to life, must be considered in deciding whether it is worth attempting to control the effects of the menstrual cycle, bearing in mind the difficulty of predicting high-risk days at the present time.

Another factor which must be taken into account in this context is the extent to which the work being carried out involves other people. There is a lot of evidence of emotional changes during the menstrual cycle (Altmann, Knowles and Bull, 1941; McCance, Luff and Widdowson, 1937). "Critical events" in the author's experiments described earlier were found to be significantly related to the menstrual cycle. These results suggest that attitudes to, and dealings with, people are particularly susceptible to the effects of the menstrual cycle. Thus considerable effects are likely to be experienced in those jobs which involve a great deal of interpersonal relationships. This is particularly important with regard to supervision. Dalton (1960a) found that school prefects were much stricter during the pre-menstrual phase of the cycle and around menstruation.

If it is necessary to control the effects of the menstrual cycle, how can it be done? In sport, cycles are shifted by hormone treatment so that the pre-menstrual and menstrual phases occur before major events (Brunelli and Rottini, 1965a; Cseffalvay, 1966). However, in work situations it is often not so much a case of trying to push the cycle so that a particular stage occurs on a particular day, but rather of predicting what stage of the cycle will be reached every day. Hormone pills make the incidence of menstruation predictable on the principle of withdrawal bleeding. In these circumstances the cycle is artificial, and, although bleeding occurs, this is not of the same type as that caused by the break up of the endo-metrium, as in normal cycles (see Appendix I). Indeed, these artificial cycles are being experienced by the very large number of women who take contraceptive pills regularly; however, no conclusions can be drawn in these cases because the effects of the pill on behaviour are not known, and there is also the possibility that, although the pill does affect some of the physiological aspects of the menstrual cycle, some of those behavioural and emotional changes which accompany the menstrual cycle still persist. These changes will be out of phase with the artificial cycle and may not be predictable. Further research in this area is needed.

An alternative way of reducing the effects of the menstrual cycle would be to employ only men on jobs where changes in capacity of the kind experienced in the menstrual cycle are likely to be crucial. This cannot really be justified, however, because men also experience changes in capacity for various reasons which may include secondary effects of the menstrual cycles of their wives. Indeed, if it were possible to forecast changes in performance due to the menstrual cycle (which could happen in the future) it might be **preferable** to employ women, because their outputs would then probably be more predictable than those of men, for whom the causes of changes in capacity are to a large extent unknown.

It seems, therefore, that at the present time there is very little which can be done to reduce the effects of the menstrual cycle. More research should be directed towards a better understanding of the cycle and its effects. A major difficulty is that women are to a considerable extent still affected by taboos and superstitions which make them reluctant to provide information, and which would probably prevent them from taking advantage of any developments. Thus a programme of education is also necessary, to enable women to understand the changes which occur during the menstrual cycle and to cope with the effects which these have on the capacity to deal with the demands of normal life. Even at the present time many women would benefit from treatment for pre-menstrual tension, but they believe that the symptoms which they experience are unavoidable concomitants of the cycle and so they seek neither treatment nor advice.

Finally, it is worth bearing in mind that the effects of the menstrual cycle are not confined to the individuals concerned, but may have significant effects on those about them. Where, for example, PMT leads to kleptomania (a not uncommon event) the stress on the rest of the family if the woman is apprehended is great, and this will undoubtedly affect their attitudes to work and to other people. Again, sometimes women at certain phases are less able to meet their domestic commitments, e.g., they may get up late, with the result that the rest of the family is late for work or school. There are indeed many ways in which the menstrual cycle can influence others, so that, when we evaluate its importance, we should consider not only its effects on individuals, but also its impact on society as a whole.

# Appendix I.

*The physiology of the menstrual cycle*

The menstrual cycle consists of the maturation of an ovarian follicle and its expulsion from the ovary, followed by the development of the corpus luteum. This process, which takes place in the ovaries, is accompanied by the growth of the endometrium in the uterus. If pregnancy does not ensue, the corpus leuteum regresses and the endometrium degenerates and is expelled ʼfrom the body with the menstrual flow. These changes are controlled by hormones released from the pituitary and ovaries during the various phases of the cycle.

The menstrual cycle is usually considered in 4 stages. The **pre-ovulatory** phase is the follicle ripening phase during which ovarian follicles ripen under the influence of follicle stimulating hormone (FSH), which is released from the pituitary. In each cycle a number of follicles develop, but all except 1 (occasionally 2) degenerate at various stages of their growth. The follicle which matures reaches the Graafian (mature) stage very rapidly in the 3 to 4 days immediately preceding ovulation. Ripening follicles produce oestrogen, which performs 3 functions at this stage of the cycle: it inhibits the release of FSH; it stimulates the release of luteinizing hormone (LH) from the pituitary; and it stimulates development of the endometrium.

The **ovulatory** phase of the cycle refers to the expulsion of the ovum from, followed by the development of the corpus luteum in the ovary. This development takes place under the influence of prolactin, which is released from the pituitary. The corpus luteum secretes progesterone, which in conjunction with some oestrogen, prepares the endometrium for the nidation of the fertilized ovum. If pregnancy does not ensue, oestrogen and progesterone inhibit the release of prolactin in the pituitary and the corpus luteum begins to regress.

The regression of the corpus luteum starts the **pre-menstrual** phase and results in the withdrawal of oestrogen and progesterone, which leads to the degeneration of the endometrium.

During the **menstrual** phase of the cycle, fragments of the endometrium are cast off and make up part of the menstrual issue. Withdrawal of oestrogen and progesterone removes the inhibitory influences on the pituitary, which once more secretes FSH, and a new cycle begins.

## References

Abramson, M. and Torghele, J. R. (1961). *Amer. J. Obstet. Gynecol.*, **81**, 223–232.
Altmann, M., Knowles, E. and Bull, H. D. (1941). *Psychosom. Med.* **3**, 199–225.
Anon. (1930). *In* "Occupation and Health" No. 152, p. 28, I.L.O. Geneva.
Arey, L. B. (1939). *Amer. J. Obstet. Gynecol.* **37**, 12–29.
Balasz, J. V. (1936). *Psychiat. Neurol. Wochenschr.* **38**, 407–409.
Benedek, T. and Rubenstein, B. B. (1939a). *Psychosom. Med.* **1**, 243–270.
Benedek, T. and Rubenstein, B. B. (1939b). *Psychosom. Med.* **1**, 461–485.
Benedict, F. G. and Finn, M. D. (1928). *Amer. J. Physiol.*, **86,**, 59–69.
Bickers, W. (1951). *Virginia J. Sci.* **2**, 210–214.
Bramson, J. (1929). *Psychiat. Neurol. Bl. Amst.* **1**, 63–76.
Bruce, J. and Russell, G. F. M. (1962). *Lancet*, **ii**, 267–270.
Brunelli, F. and Rottini, E. (1965a) *Medna Sport*, **5**, 832–841.
Brunelli, F. and Rottini, E. (1965b). *Medna Sport* **5**, 822–831.
Cooke, W. R. (1945). *Amer. J. Obstet. Gynecol.* **49**, 457–472.
Cseffalvay, T. (1966). *Med. Sport, Berl.* **6**, 1–5.
Dalton, K. (1954). *Brit. Med. J.* **ii**, 1071.
Dalton, K. (1960a). *Brit. Med. J.* **ii**, 326–238.
Dalton, K. (1960b) *Brit. Med. J.* **ii**, 1425–1426.
Dalton, K. (1961). *Brit. Med. J.* **ii**, 1753.
Dalton, K. (1964). "The pre-menstrual syndrome". Heinemann Medical Publ. Co., London.
Dalton, K. (1968). *Lancet*, **ii**, 1386–1388.
Eagelson, H. E. (1927). *Comp. Psychol. Monogr.* **4**, (20) p. 65.
Eayrs, J. T. and Glass, A. (1962). *In* "The Ovary" (S. Zuckermann, ed.) Vol. II. Academic Press, London and New York.
Farris, E. J. (1956). "Human Ovulation and Fertility". Pitman Medical Publ. Co., New York.
Fichera, C. and Romano, S. (1965). *Medna Sport* **5**, 406–411.
Fluhmann, F. (1956). "The Management of Menstrual Disorders". W. B. Saunders and Co., Philadelphia.
Frank, R. T. (1931). *Arch. Neurol. Psychiat.* **26**, 1053.
Freed, S. C. (1945). *J. Amer. Med. Ass.*, **127**, 377–379.
Gilman, J. (1942). *J. Clin. Endocrin.* **3**, 157–160.
Glanville, E. V. and Kaplan, A. P. (1965). *Nature, (London)* **205**, 930–931.
Gorkine, Z. D. and Brandis, S. (1936). *Arbeitsphysiologie*, **9**, 254–266.
Govorukhina, E. M. (1964). *Akush. Ginekol. (Moscow)* **3**, 95–101.
Greene, R. and Dalton, K. (1953). *Brit. Med. J.* **i**, 1007–1014.
Greenhill, J. P. and Freed, S. C. (1940). *Endrocrinology*, **26**, 529–530.
Gunn, D. L. (1937). *J. Obstet Gynaecol. Brit. Emp.* **44**, 839–879.
Harvey, O. L. and Crockett, T. (1932). *Hum. Biol.* **4**, 453–468.
Herren, R. Y. (1933). *J. Exp. Psychol.* **16**, 324–327.
Hitchcock, F. A. and Wardwell, F. R. (1929). *J. Nutr.* **2**, 203–215.
Holt, J. G. J. (1935). *Zentralbl. Gynaekol.* **59**, 1161–1164.
Israel, H. (1938). *J. Amer. Med. Ass.* **110**, 1721–1723.
Johnson, G. B. (1932). *J. Comp. Psychol.* **13,**, 133–141.
Kenshalo, D. R. (1966). *J. App. Physiol.* **21**, 1031–1039.
Kirihara, H. (1932). *Rep. Inst. Sci. Labour, Kurasilki*, **14**.
Kleitman, N. (1949). *Physiol. Rev.* **29**, 1–30.

240 JUNE A. REDGROVE

Kroger, W. S. and Freed, S. C. (1951). "Psychosomatic Gynaecology". W. B. Saunders and Co., Philadelphia.
Lamb, W. M. (1953). *Amer. J. Psychiat.* **109**, 840–848.
Lederer, J. (1963). *Ann. Endocrinol.* **24**, 460–465.
Lewin, K. and Freund, A. (1930). *Psychol. Forsch.* **13**, 198–217.
Loeser, A. S. (1943). *Lancet*, **1**, 518–519.
Lough, O. M. (1937). *J. Genet. Psychol.* **50**, 307–322.
Mackinnon, I. L. (1954). *J. Obstet Gynaecol., Brit. Emp.*, **61**, 109–112.
Marshall, J. (1963). *Brit. Med. J.* **i**, 102–104.
McCance, R. A., Luff, M. C. and Widdowson, E. E. (1937). *J. Hyg.* **37**, 571–611.
Menaker, M. (1959). *Amer. J. Obstet. Gynecol.* **77**, 905–914.
Middleton, W. C. (1934). *Psychol. Clin.* **22**, 232–247.
Moore, L. M. and Cooper, C. R. (1923). *Amer. J. Physiol.* **64**, 416.
Morton, J. H. (1950). *Amer. J. Obstet. Gynecol.* **60**, 343–352.
Morton, J. H. (1953). *Amer. J. Obstet. Gynecol.* ,**65** 1182–1191.
Pierson, W. R. and Lockhart, A. (1963). *Brit. Med. J.* **i**, 796–797.
Rees, L. (1953). *J. Ment. Sci.* **99**, 62–73.
Redgrove, J. A. (1968). "Work and the Menstrual Cycle". Unpubl. Ph.D. thesis, University of Birmingham.
Reich, M. (1962). *Australas. Ann. Med.* **11**, 41–49.
Reynolds, S. R. M. (1942). *Amer. J. Obstet. Gynecol.* **44**, 155.
Rodriques da Costa Doria, J. (1947). *Impr. Med. Rio de J.* **22**, 25–26.
Rosenzweig, S. (1943). *J. Clin. Endocrinol.* **3**, 296–300.
Rubenstein, B. B. (1938). *Endocrinology*, **22**, 41–44.
Schwarz, W. (1959). *Diss. Abstr.* **19**, 3372.
Seiichi, M. (1962). *Gumma J. Med. Sci.* **11**, 294–318.
Seward, G. H. (1934). *Psychol. Bull.* **31**, 153–192.
Sfogliano, C. (1964). *Rass. Med. Ind.* **33**, 218–221.
Smith, A. J. (1950a). *J. Appl. Psychol.* **34**, 1–5.
Smith, A. J. (1950b). *J. Appl. Psychol.* **34**, 148–152.
Sowton, S. C. M. and Myers, C. S. (1928). *Ind. Fatigue Res. Bd.* Rep. No. 45.
Tinklepaugh, O. L. and Mitchell, M. B. (1939). *J. Genet. Psychol.* **54**, 3–16.
Torghele, J. R. (1957). *J. Lancet.* **77**, 163–170.
Van de Velde, T. H. (1939). "Über den Zusammenhang zwischen Ovarialfunction, Wellenbewegung, und Menstrual Blutung, und über die Entehung des sogenannten Mittelschmerzes". Haarlem: Bonn.
Vernon, P. E. and Parry, J. B. (1949). "Personnel Selection in the British Forces". University of London.
Vierling, J. S. and Rock, J. (1967). *J. Appl. Physiol.* **22**, 311–315.
Wickham, M. (1958). *Brit. J. Psychol.* **49**, 34–41.

# CHAPTER 7

# Industrial Work Rhythms

K. F. H. MURRELL

*Applied Psychology Department, University of Wales Institute of Science and Technology, Cardiff, Wales*

## I. Work Curves 1920–1953

Rhythms which will be dealt with in this chapter differ from some of those discussed elsewhere in that, generally, they are not demonstrated by measurements which can be made of processes within the individual, as, for example, is the case with temperature in the circadian rhythm. Rather, the changes which take place throughout a working day will be most often shown by changes in the rate or nature of work, typically when this work is repetitive. This does not mean that diurnal changes in work pattern are not determined by changes within the individual, because this is almost certainly the case, but that there is difficulty in disentangling the effects which are due to the individual and those which may be due to the circumstances in which the individual works. For instance, in work which is moderately heavy a fall in output may be due to physical fatigue, but on the other hand it may equally well be due to the method of payment in that an operative may have approached his output target half way through the working day and therefore have decreased his rate of work for the remainder of that day.

Observations of changes in work rate are not easy to obtain since they involve continuous recording of rate of production throughout the working day for many days on end; it is not surprising therefore that the number of studies in this area is relatively few. Most of these studies have been undertaken with the object of investigating either fatigue, or monotony and boredom, but other factors which influence the shape of the work curve, including pacing (either by machine or by other workers), the size of the batch, the length of time the operative has been on the job and the system of payment have also been investigated at one time or another.

The majority of the investigations into fluctuations of production rate with time were carried out between the wars. There was a small resurgence of interest in the late 1940s and early 1950s but such research seems virtually to have ceased from then until quite recently. In fact when work curves are discussed in relation to such factors as fatigue, monotony or boredom even the most recent books on industrial psychology quote this inter-war work with few, if any, references to studies later than 1955. This early material is relatively sparse in the kind of work curve which we require for our purpose, but this is not very surprising since it is only recently that the development of electronic work recording has made it possible to do extensive week-long studies both accurately and economically.

The literature on fluctuations in output during longish periods of work such as might occur in industry is confused because of the use of the term "work curves" when describing **exercise** over quite short periods. Two such studies emanated from Poffenberger's laboratory. A study of "variability of performance in the work curve" by Weiland (1927), although published in the *Archives of Psychology*, turns out to be a study of short periods of muscular work, using various types of dynamometer or ergograph. A similar study in the same journal by Manzer (1927) on "rest pauses" is likewise a physiological study in which rests of varying duration were given. Researchers in the 1920s and 1930s seem to have been preoccupied with laboratory activity over comparatively short periods of time (e.g. Bills, 1931 or Phillpott, 1932) and there are relatively few reported studies of fluctuations in work performance over as long as half a day, either in the laboratory or in the real life situation. However, some of these early long term studies are of interest. For example, in the laboratory Arai (1912) observed his own performance when multiplying, mentally, 4 digits by 4 digits continuously for about 12 h per day on 4 successive days. The rate of multiplication towards the end of each day was about half of what it had been soon after the beginning of the session of work (this experiment has been repeated with a larger number of subjects by Huxtable, White and McCartor (1946), who got a similar result). Poffenberger (1928), whose work is referred to in more detail below, found that after 5 h of intense work at addition the rate of work was reduced by about 20%. Reed (1924) made his subjects do mental addition for a 10 h day with $\frac{1}{2}$ h for lunch and found there were only slight decreases in rate of work after this period.

One group which carried out extensive shop floor studies during the period under review was the Industrial Fatigue Research Board of the Medical Research Council (later renamed the Industrial Health Research Board). However, their early work was not so much concerned with fluctuations in output as with the effect of introducing a rest for refreshment in the middle of a long period of work. For instance, in a study by Wyatt and Fraser (1925) the mean time to fold and box one dozen handkerchiefs was taken over each 15 min of the working day, for 8 girls, for 3 weeks, under a condition of continuous work and also when a break was introduced after $2\frac{1}{4}$ h. Although it was not the object of the exercise, what is interesting about this study is the very wide variety of individual production curves which was obtained, 2 showing a sharp fall after $2\frac{1}{4}$ h work (Fig. 1), 1 curve having 2 sharp peaks of output, 4 showing a steady decline and 1 a steady increase. The authors used these curves to support the idea that a rest should be given after $2\frac{1}{4}$ h work, but there is little evidence of consistency in the individual work rhythms.

These curves are in any case somewhat at variance with Wyatt's earlier work, since he had previously shown that when output is observed over a long period (in his case 6 months) the resulting data suggest that industrial workers unconsciously succeed in maintaining a steady production rate throughout the working hours (which at this time were 44–48 h a week). In this study (Wyatt, 1923) the hourly output of groups of 7 to 25 weavers working in 5 different sheds was recorded and it was found that their output did not vary from the mean by more than 3% throughout the working day except for a rise during the first $\frac{1}{2}$ h, which is stated to be due to the "warming up" effect. It is true that weaving is a "machine minding" job whereas handkerchief folding is not, but another typical study by Vernon, Bedford and Wyatt (1924) of an individually self-paced job gave similar results. They studied the labelling of small packages, an operation which had a cycle time of about 4 sec. The production of 7 girls was measured in 3 separate weeks for 4 days, Monday morning and Friday afternoon being omitted. During the first week no rest was given and in the immediately following week a rest pause of 10 min was introduced. After 2 weeks on this schedule a second set of observations was taken and a third set after 11 weeks (Fig. 2). In the first 2 periods the output remained fairly steady throughout the day but in the third period it fell off very rapidly towards the end of the working time. The authors say that this was not due to the introduction of the rest pause but to a sudden burst of very hot weather, which made the girls unable to keep a high rate of performance going all day. Commenting on this Vernon

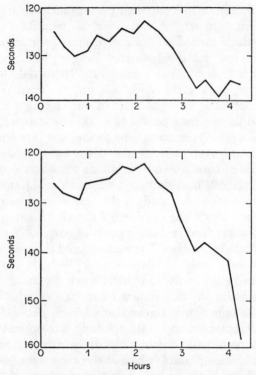

FIG. 1. Time taken by 2 operators to fold 1 dozen handkerchiefs during morning work period (after Wyatt and Fraser, 1925).

(1924) says "these results show how unreliable hourly output records may be unless they are continued for such a period as to neutralize the effects of temporary changes in the conditions of work". The fairly smooth curves obtained in these and similar studies will be shown only when output is aggregated over a very long period from a fairly large number of workers over a fairly long unit of time. In emphasizing the way in which individuals depart from these idealized curves, Vernon gives as an illustration the daily output of 1 of his 7 labellers. This is reproduced in Fig. 3.

In spite of this and other similar studies we find Burtt in 1929 producing a smooth curve showing a performance decrement from about the third hour of the work (Fig. 4). Burtt says that this decrement "is probably traceable to fatigue". He shows also the warm-up effect, quoting Robinson and Heron (1924). A similar curve appears in Katzell (1950) (Fig. 5.B), which in turn is reproduced in Siegel (1962) (Fig. 6) and no doubt in

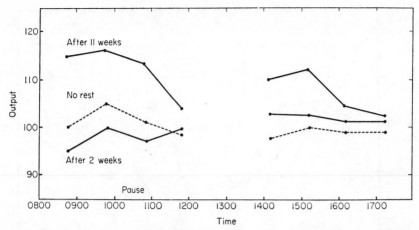

FIG. 2. Mean output of 7 girls labelling packages (after Vernon, 1924)

other text books on Industrial Psychology as well. There has been a tendency to describe this curve as a "fatigue curve".

Although it is probable that there are work rhythms in activities which are non-repetitive such as machine minding, it is in jobs which are mainly light and in which the same movements are repeated over and over again that changes with time (if they exist) can best be detected. Since this work is repetitive it is basically monotonous and it has been assumed that in this situation there are 2 main factors which may cause fluctuations in the rate of work: "boredom" and "fatigue". Baldamus (1951) has suggested that the term "boredom" may be confusing since it may subsume several different phenomena. He suggested that boredom with the subject matter or content of the work should be distinguished from the effect of repetition, which he called "tedium". He illustrates this by suggesting that a book may be extremely boring without necessarily being repetitious. Be this as it may, the term boredom is most commonly used in relation to repetitive work to describe the phenomenon for which Baldamus has proposed the term tedium; for this reason it will be used in this sense throughout this chapter. "Fatigue" is much less easy to pin down. This is partly historic since in the past, when most work was heavy, a worker became physically tired and therefore the term "fatigue" had some real meaning; when incentive payments were first introduced an allowance for this fatigue was given which has been named a "fatigue allowance", "compensating rest" or "relaxation allowance". However, as work has become lighter workers no longer get physically tired in the sense that they did when they were acting as sherpas; there has however been no corresponding change in thinking, so that allowances are now given for a

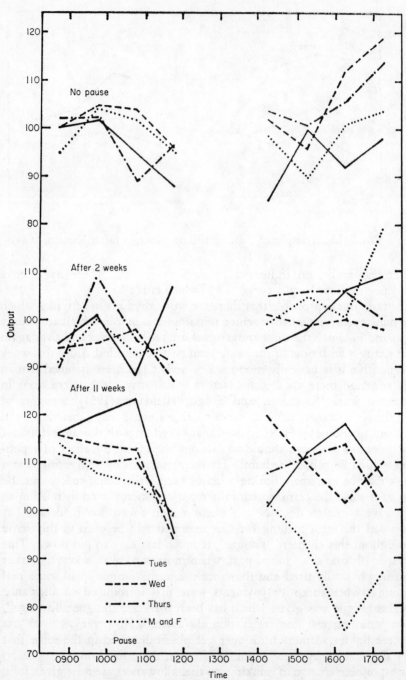

FIG. 3. Output of a labeller (after Vernon, 1924).

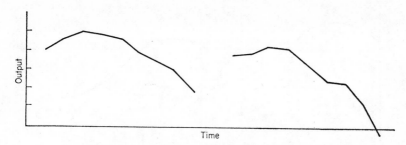

FIG. 4. Theoretical "fatigue" curve (after Burtt, 1929).

phenomenon which is not understood and which cannot be measured (it is interesting to note that the virtual disappearance of heavy physical work in most manufacturing industry coincided with the development of accurate methods of its measurement by physiological means, but that these have not, in turn, been replaced by methods of measurement of non-physical work as the nature of work has changed).

Both "boredom" and "fatigue" have been inferred from association of the shape of production curves with the assessment of the feelings of

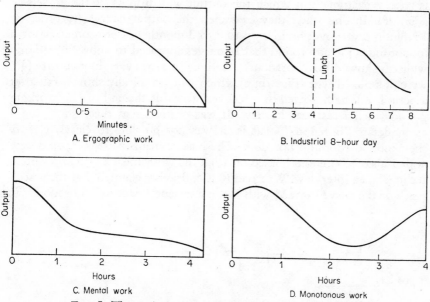

FIG. 5. Theoretical output curves (after Katzell, 1950).

I

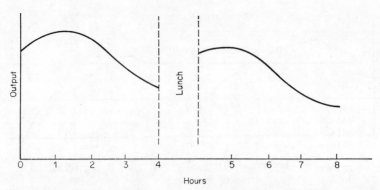

Hours

FIG. 6. Theoretical output curve (after Siegel, 1962).

workers as expressed in interviews or questionnaires. The earliest systematic work in this area which can be traced is that of Wyatt and his collaborators (Wyatt, 1929; Wyatt *et al.*, 1937) who carried out investigations for the Industrial Health Research Board. These workers proposed that the nature of the effects could be inferred from the shape of the work curve. They suggested that when "boredom" is present the output will increase throughout the working period because the end of the span of monotonous work is approaching (Fig. 7d). The opposite (they suggest) is true of "fatigue", in which the output will drop as "fatigue" sets in (Fig. 7a). In one study they examined the output curves (recorded at 15 min intervals over periods up to 2 to 3 months) of 68 women doing a monotonous job and found that 16 curves appeared to show "boredom" whilst 6 showed no "boredom" ("fatigue"). However, the majority (46) were of a mixed type (Fig. 7b, c) which would typically show a fall after a period of time followed by a recovery towards the end of the working period when a rest was in view. It was claimed that these show slight to moderate "boredom", that is, it was not possible to clearly classify them into boredom/no boredom. These assessments were compared with appraisals of "boredom" based on what the women have said about their feelings in an interview. Wyatt and his colleagues claimed that the assessment of the curves agreed with the assessments made of "boredom" in

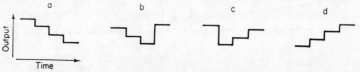

FIG. 7. Theoretical output curves (after Wyatt *et al.*, 1937).

51 out of the 68 women who were examined, which is far higher than would have been expected by chance alone.

Unfortunately Wyatt does not give any example of the actual output curves on which he based his theoretical curves shown in Fig. 7. He does however give output curves derived in a subsequent experiment; one of these is illustrated in Fig. 8. If these are typical of the kind of output

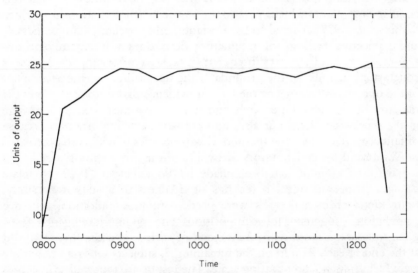

FIG. 8. Average output per quarter-hour, morning period (after Wyatt *et al.*, 1937).

curves which these researchers had been obtaining in the 3 factories in which they were working then it is difficult to see how they could be classified into any 1 of the 4 types with any degree of certainty. It is not surprising, therefore, that only 22 out of the 68 curves could in fact be classified clearly as showing either "boredom" or "no boredom" and that only 14 of these coincided with the assessed feelings of the workers involved. Nevertheless the authors claimed that there was a high relationship between the feeling of "boredom" and the shape of the output curve. It should be noted that the theoretical "fatigue" curve differs from that originated by Burtt (1929) in that no warm-up is shown (Wyatt mentions warm-up in another section of his report). Katzell (1950) produces a curve purporting to show "boredom" (Fig. 5d), but without giving any authority; and this curve is entirely different from that proposed by Wyatt. Despite these conflicting curve shapes this work has given rise to the idea that standard work curves can be drawn to represent "fatigue" or "boredom", and we find this concept being repeated even very recently in various textbooks on working efficiency.

There are however a number of objections to the work on which this concept is based. Firstly, as Murrell (1962) has pointed out, the state of "fatigue" or what-have-you must be inferred from changes in the shape of a production curve; this "fatigue" is then given as the reason for the changes in the curve from which it was originally inferred. This is a circular argument which gets one nowhere (the relationship which Wyatt et al. (1937) claimed to have established between a "fatigue curve" and subjective feelings of fatigue/boredom does not circumvent this argument). Secondly, most subsequent workers have not supported Wyatt's conclusions, mainly on the grounds of the unreliability of the assessment of the operatives feelings or the lack of evidence that output is in fact the measure of the same type of change as might be measured by a change in the workers feeling, but also, and perhaps most importantly, because of failure to reproduce the theoretical work curves which were postulated by Wyatt and his collaborators or by any of the other authors cited.

Earlier an attempt had been made by Poffenberger (1928) to relate changes in performance to feelings of tiredness in a laboratory study. Four kinds of mental tasks were used: continuous addition, sentence completion, judgement of compositions and taking intelligence tests. Each task lasted about $5\frac{1}{2}$ h. At the end of each $\frac{1}{2}$ h on the first task, and at the end of each 21 min on the remaining 7, subjects reported how they were feeling on a 7 point rating scale. The results are reported as averages of 10 to 13 subjects, from a total of 14 (see Fig. 9). From these group results there appears to be no consistent relationship between the output and a feeling of tiredness (although, according to the author, there is in some individual cases a suggestion that "those who showed the greatest fall in output also show the greatest change in feelings."). It will be seen that, in the case of the intelligence test, performance improved steadily over the work period until it was nearly 20% better than at the outset, whereas the reports of feeling showed an increase of tiredness of 1·3 points. On the other hand performance in addition deteriorated after about the first 2 h of work, while the feeling score also deteriorated. It is true that this is a laboratory study of work undertaken, presumably, by "amateur" subjects; it may therefore not necessarily represent the situation existing when people are "professionally" involved in the task which they are performing. Although this work is directly relevant to their investigation Wyatt et al. (1937) do not refer to it. It is clear that it throws grave doubts on the validity of Wyatt's attempts to relate curve shapes to subjective feelings.

Roethlisberger and Dickson (1941) also published results which did not

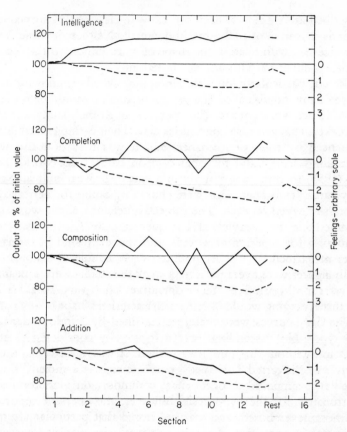

FIG. 9. Relation between output and feelings in 4 kinds of mental work
(after Poffenberger, 1928).

confirm the Industrial Health Research Board findings. This was an
industrial study and (except in the case of one operator) did not produce
a work curve which in any way resembled that which is claimed to be
typical of a "boring" situation, in spite of the fact that the work was both
repetitive and monotonous. A similar study by Rothe (1946) showed
that the daily work curves of different individuals varied very widely and
did "not assume any characteristic predictable pattern". Correlations
between the work curves of the same individual on different days also
varied widely, some subjects showing a relationship, others none whatsoever. But when Rothe averaged the work curves of each of his workers
over one week he obtained curves from 2 subjects which he classified as
"fatigue curves", two which he called "monotony curves", and four which

1§

he described as "mixed". These curves were classified by inspection and
are clearly a matter of individual judgement. Neither of these 2 studies
were concerned with relating the shape of work curves to the feelings of
the workers either of "fatigue" or "boredom".

Rothe was particularly interested in finding whether, knowing the type
of curve of any individual or of a group, predictions could be made of the
shape of future work curves. The operators in Rothe's study were hourly
paid workers engaged on a continuous task. Their earnings were therefore
uninfluenced by their performance, and it was thus doubtful whether
such findings were valid for operators on batch piece work. This led
Smith (1953) to carry out a study in a small knitwear mill in Northern
Pennsylvania. Two operations were chosen by Smith for her study, with
8 workers engaged on each. The subjective feelings of the workers were
determined both by interview and by questionnaire. Output was recorded
continuously for a period of a week and was expressed in terms of the
number of garments sewn per minute over a bundle of 5 dozen. When
the daily figures were averaged over a week the results were a meaningless
flat curve for all operatives. Since operatives often vary widely from day
to day this averaging would obscure a characteristic shape of any particular
day. The daily curves were therefore classified by 2 independent judges
into 4 types. No "ascending" curve (classically considered to indicate
"boredom" without "fatigue") was found. It was agreed that there was
possibly one "inverted U-shaped" curve, which is a mixed "boredom/
fatigue" type supposed to occur most commonly on monotonous tasks;
"unfortunately for the hypothesis, however, the operator reported she
was almost never bored, and was not bored that particular afternoon".
Approximately one third of the curves were classified as "descending"
(supposed to indicate "fatigue") but the judges agreed on less than half
of these. The remainder of the curves were considered unclassifiable by
both judges. It was not possible therefore to relate the shape of the
work curves found to any results obtained from the questionnaire about
the operatives' feelings. In her paper Smith gives the output curves for a
complete week for 1 operator and these are reproduced in Fig. 10.

This work shows clearly that one of the major problems in the use of
output curves in research is the difficulty of classification, as Smith (1953)
has pointed out. There are no satisfactory ways of describing in detail
the shapes of these curves, nor are there satisfactory statistical measures
available for comparisons. Correlation techniques can be used but the
resulting coefficients cannot allow for overall similarity in the shape of
curves when the changes of slope are displaced slightly in time, nor do
they allow for peculiarities of distributions. The only alternative is to use
visual inspection, which is subjective and unreliable. Daily fluctuations

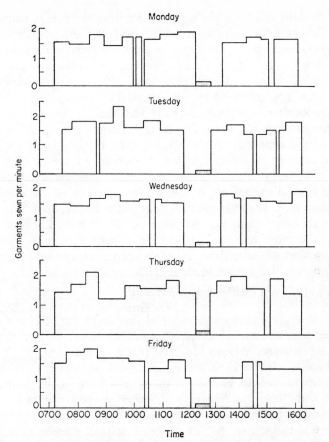

FIG. 10. Output curves for 1 week for operative hemming garments (after Smith, 1953).

can make weekly production curves flat and almost meaningless, and even when day-to-day curves are assessed there is no agreement between subjective and statistical assessments of the shape of the function. Nor is there likely to be agreement between judges if the experience of Wyatt *et al.* (1937) is any guide.

It is difficult now to realize what work conditions were like in the 1920s and 1930s. For instance Hersey (1925) writes about the coil winding department of a textile mill where 50 young girls (aged 13–22) worked for 9¾ h a day. The foreman complained that the "sneak to the toilet" was the bane of his existence. "So prevalent was the trouble that he could pick out no one person not guilty. Recognizing that the length of working period justified some trips to the dressing room (sic), he could devise no standard rule to apply without incurring the risk of encroaching too heavily upon his workers' actual rights". A further example which Hersey gives is of 5 girls, working in a plant where no authorized resting was permitted, who admitted that where the forewoman would not "let them sneak off" they found a way to put their machines temporarily out of order. In another plant where the men worked a 10-h day they would leave their machines after working for 2 or $2\frac{1}{2}$ h and spend 15 min to 20 min in the only restroom available, the toilet. In the afternoon the same procedure was followed except they would lower a wire to a boy from a shop and bring up some food for a little lunch. Conditions are of course very different now, and this may be a reason for doubting the applicability, in present industrial conditions, of the work quoted so far. The report by Smith mentioned above seems to be the last piece of work in this area until very recently; at any rate no other published work has been traced, and even the most recent textbooks on industrial psychology do not refer to any work of a later date. Smith's work may therefore be taken as the end of a phase which started in about 1920 and lasted until 1953; it represents an appropriate point for summing up.

In spite of the early work which suggested that in some instances there might be no work decrement at all, and that there was so much within-worker and between-worker variability that the course of output could not have been predicted, various idealized forms of industrial work curves had been established by 1930. Subsequent work attempted to relate these work curves to various factors in the work situation, in particular to "fatigue" and "boredom"; later still, other research showed that this kind of classification was not viable. So, in a sense, the wheel has gone full circle and the subject is back where it started; there may indeed be fluctuations in output with time but the reason for these fluctuations is still obscure. It is true that there will often be organizational reasons for hourly or daily changes in output; production targets may already have been met, there may be machine breakdowns, shortages of material, high or low temperature or humidity. But even when conditions are ideal there may still be rhythms within the worker himself which may influence the rate of production to cause decrements towards the end of the working period. It was in order to explore this possibility that Murrell in 1962 started a programme of research into repetitive and monotonous work.

## II. Laboratory Research into Work Rhythms

This work started from a desire to "debunk" the idea of "fatigue". The motivation for this lay in the prevalence of extensive fatigue allowances being given in industry for work which was light and which could not

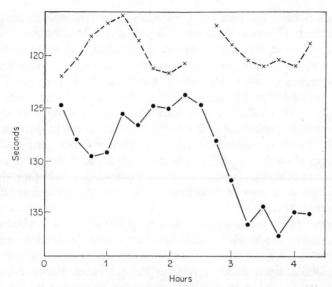

FIG. 11. Output (as measured in time taken per dozen) of an operative folding handkerchiefs under 2 conditions: top, with rest; bottom, without rest (after Wyatt and Fraser, 1925).

therefore be said to be "fatiguing" in the physical sense. If it is borne in mind that these allowances constitute about 15% of the total wages bill the importance of establishing the true nature of the phenomenon for which the allowances were being given will be appreciated. The starting point for the work was the assumption that when work is light and repetitive fluctuations in the rate of work will be due to mental rather than physical causes. It was argued (Murrell, 1962) that the early results of the Industrial Fatigue Research Board investigators (e.g. Wyatt and Fraser, 1925) had repeatedly shown that when rest was introduced into a long period of continuous work, output went up not only **after** the rest but also **before** it (Fig. 11). The investigators suggested more than once that this was due to the workers not having to look forward to long periods of uninterrupted work.

If it is accepted that under modern conditions any changes in production which may occur have some mental, rather than physical basis, then we can look to the literature on psychological fatigue for suggestions as to the possible causes of any fluctuations which may be observed. Bartlett (1953) suggested that when work was repetitive the onset of psychological "fatigue" would be shown by the appearance of irregularity in the execution of successive repetitive movements. This should be accompanied by a change in the standard deviation of successive cycle times, and Murrell

(1962) examined this possibility in relation to data which he obtained from Industrial Fatigue Research Board Reports. It did not seem feasible to develop a method based on standard deviation, and so instead he proposed a method based on quality control techniques. It is characteristic of most repetitive work that after a warm-up there is a period of fairly regular and stable performance. This period has been named the "actile" period (Murrell, 1962). The proposed techniques involved identifying the actile period and calculating its mean and standard deviation, then setting limit lines at two standard deviations from this mean. Any cycle which was longer than the upper limit was counted and called a "long time". It was to check the viability of this technique and whether there was actual physical fatigue in repetitive work that Murrell's first laboratory experiment was set up.

Vernon (1924) quotes an experiment by Burnett (1924) who simulated an industrial situation with 4 subjects who worked for $2 \times 3$-h spells a day for 4 days a week for 8 weeks. She got results which were entirely different from those obtained under similar experimental conditions on the shop floor. Vernon goes on to say "the conclusion suggested by the above is obvious. In order to obtain information which can be directly applied to industrial problems it is necessary to collect this information strictly under industrial conditions". There is unfortunately a good deal of truth in this assertion, but circumstances do arise when the only way of making progress is to go into the laboratory. That, in the event, Murrell's results did not prove to be directly applicable in industry (as will be described later) does not mean that they are invalid.

The task developed was a simulation of the testing of electrical components, an operation which is found in most electronic manufacturing factories. The component is put into a jig, buttons are pressed, a meter is read and the component is disposed of according to the readings. In the simulated task ball bearings were put into 6 holes and a key was then pressed which allowed the index of a meter to swing to 1 of 5 positions. According to the position at which the needle stopped, a coded response involving 2 out of 4 keys was made. If this had been done correctly a green light came on, another key was pressed and the ball bearings fell back into a tray. The cycle time for this operation varied between about $7\frac{1}{2}$ sec at the outset to about 5 sec or less after 2 or 3 months practice. Subjects were hired for a period of 3 to 4 months and were paid a fixed weekly wage, except when the effect of incentives was one of the objectives of the experiment. They spent 4 h in the laboratory, of which $3\frac{1}{2}$ h was spent at work either in the morning or the afternoon. A number of working conditions were tested, but from the seventh subject onwards work on Tuesday was always done without a rest as a calibration day (up to

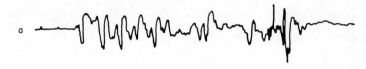

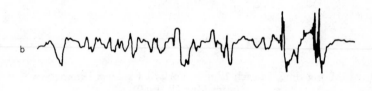

FIG. 12. Electromyographs: (a) after 15 min work and (b) after $3\frac{1}{2}$ h work (Subject A); (c) after $3\frac{1}{2}$ h work and (d) after cranking to exhaustion (Subject B).

this point this calibration session had been held on different days of the week but a strong suggestion of a day-of-the week effect had been noticed, which, if present, would have made it difficult to fit the learning curve necessary if valid statistical tests were to be made on the data).

In the first experiment (Murrell and Forsaith, 1963) electrodes were attached over the extensor digitorum of the preferred forearm of the subject. Electromyographic readings were taken at the beginning, in the middle and at the end of $3\frac{1}{2}$ h periods of work. There was virtually no change in the electromyographic trace. The subjects were then asked to crank to exhaustion against a load and the repetitive movements were

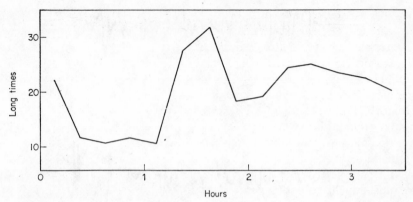

FIG. 13. Long-times in each 15 min work of 1 subject summed over 10 days
(after Murrell, 1969).

then resumed. The difference between the traces at the end of the $3\frac{1}{2}$ h work and those when it was known that the muscles were fatigued was very marked indeed (Fig. 12). It seemed clear, then, that any changes which took place in the rate of production could not have been due to extreme physical fatigue in the arm being used. When the long-time scores were examined it was found that there was a very sharp increase in number at about the end of the first hour to $1\frac{1}{2}$ h of work (Fig. 13). This effect varied from day-to-day and between subject and subject, but when the results were aggregated the resulting curves typically showed a very sharp increase in the number of long-times in the fifth $\frac{1}{4}$-h period. This was as had been predicted, but what had not been predicted was an improvement in regularity after a further $\frac{1}{2}$ h work. This was followed in turn by a further increase in the number of long-times towards the end of the period of work. Over a period of 6 years, 22 subjects were tested and they showed this effect fairly consistently. This suggests that the timing of the onset of irregularity may be related to the nature of the job and will not normally show a big inter-subject variation.

Thus this work shows that the irregularities predicted by Bartlett do occur after a period of time; however, any explanation for these irregularities must take into account the subsequent re-establishment of regularity, which had not been predicted. In the theoretical discussion of causes of fluctuations of performance in continuous and monotonous work, repetitive and active work appears to have been almost entirely ignored. On the other hand inactive work has been extensively studied and has been widely discussed theoretically. There seems to be no reason to suppose that the basic causes of performance fluctuations are likely to be any different for active or inactive work. If this argument be accepted then the theoretical discussion of the causes of changes in performance in, for instance, vigilance tasks may be equally applicable to the active repetitive type of task studied by Murrell. Clearly not all the explanations for changes in vigilance performance can be applied to repetitive tasks; these would include the expectancy hypothesis (e.g. Baker, 1959) or the various statistical strategies which have been suggested (e.g. Broadbent and Gregory, 1963). There is, however, one approach which does seem to hold promise; this is a theory based on arousal. It is true, as Jerison (1967) has pointed out, that almost any behaviour pattern or phenomenon can be made to fit some theory of arousal. But if we accept that behaviour changes are determined by changes in the central nervous system then we must equally accept that something akin to arousal is deeply involved.

It can be argued that the individual working in a continuous and monotonous situation will gradually become habituated to such external stimulation as there may be. This in turn will result in a reduction of stimulation of the reticular formation by projections from the afferent impulses (Hebb's cue functions). As a result the reticular activity will be reduced and the resulting performance will be steadily impaired. But we have already seen that this does not happen; work performance may, in fact, continue at a high level for a period of 1 h or more. Something more than this simple explanation is therefore required. Murrell (1967) suggested that a subject could maintain a high arousal level voluntarily by downward projections to the reticular formation as had been demonstrated by French, Hernández-Péon and Livingston (1955) in the monkey. This mechanism Murrell called "auto-arousal", which he defined as "cortical activation resulting from stimulation of the reticular formation by the cortex, this stimulation being under voluntary control". Such auto-arousal would enable the subject to maintain or even raise his arousal level at will, and this obviously opens up great possibilities for manipulating the phenomenon previously described as "fatigue".

As a first explanation of the phenomenon of the appearance of irregularity followed by its disappearance, Murrell (1966) and Bergum (1966) put forward independently the suggestion that the effect of auto-arousal would be to produce the classical "inverted-U" shaped arousal curve. It was supposed that auto-arousal would tend to produce a state of hyperarousal which would cause performance to become irregular, but that when such a state occurred a mechanism would come into play which would have the effect of damping down the activity of the reticular formation and thus cause the arousal level to fall again to a point where performance would revert to regularity. It was postulated that arousal would then continue to fall gradually until it reached so low a level that irregularity in performance would once more appear.

Unfortunately for this rather interesting theory there is evidence which suggests that it may not be valid. In the first place hyper-arousal is normally associated with what are usually called "emergency conditions" when the organism is in a very high state of stress, which is certainly not true of the subjects in the experiment. Secondly, if skin conductance can be accepted as some indication of the arousal level, on those occasions when an "inverted-U" shaped conductance curve was observed the performance irregularity did not coincide with the rising portion of the curve before it reached the summit of the "U". Rather, the irregularity occurred when conductance was starting to fall, even when there had been no U-shaped curve at all. This led Murrell (1969) to put forward the suggestion that, since the auto-arousal function might well be enhanced by the rise in general, non-specific neural activity occasional by its own operation, the damping-down mechanism would in any case have to come into play to prevent excessive arousal. If these damping-down projections could interfere with the upward and downward arousal functions it is possible that they could also interfere with the efferent activity to produce the irregularity which has been observed.

Murrell proposed that this effect might be described as "anti-arousal". As the general arousal activity is reduced the anti-arousal will also be reduced; this in turn will cause a reduction in the interference with the efferent activity, so the performance will become regular once again. By this time some 2 h may have passed and, if the skin conductance curves are anything to go by, there is generally a steady fall in arousal from this point onwards. It may be that the "motivation" which had stimulated the auto-arousal can no longer be maintained and this could mean that the gradual increase in irregularity of the performance is associated with the gradual lowering of the arousal. An alternative explanation, which is perhaps less plausible, is that once anti-arousal has been established auto-arousal can no longer function; but there is evidence which suggests that this is not the case. For instance, if after about 2 h of work a subject be given a cup of coffee or tea to drink both the arousal level and the performance immediately start to rise, which suggests that the auto-arousal function has come into play once again.

Thus we see that in the continuous and monotonous work of the kind described in the early studies of the Industrial Fatigue Research Board, it could be that there is a rhythm of regularity-irregularity which is predicted by the arousal function of the individual concerned.

Having got this far it was clearly necessary to establish whether the phenomenon which had been observed in the laboratory could also be found on the shop floor. In order to do this it was necessary to record each successive cycle time from groups of workers over a long period of

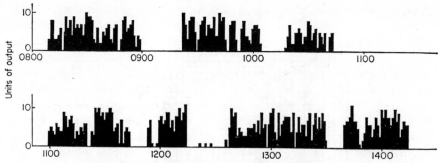

FIG. 14. Output per minute of 1 operative over 1 day.

time. Equipment for doing this was designed and built in Murrell's laboratory (it is now on the market commercially) which enabled such cycle time data to be analysed on a computer. Unfortunately, studies with this recorder failed to confirm the laboratory results. The reason for this soon became evident; it was clear that operatives never worked for long enough at a monotonous job for the irregularities to show. In fact in most of the studies the operatives concerned had become so skilled that they had to "lose" a substantial part of the working day if their output was not to exceed that expected by the management on the basis of their work study standards. Furthermore, discipline in industry has very much changed since the early days. Workers are now no longer expected to spend long periods at the bench without relief, but are generally given something like 15% fatigue allowance; this means that they can stop work for some $12\frac{1}{2}$ min in every hour. These stoppages are usually permitted by the management (in fact they would have a great difficulty in preventing workers stopping whenever they wanted to). A typical result of a day's recording is given in Fig. 14, which shows that when the girl **was** working there was no change in the rate of work from the beginning of the day to the end of the day. However, when the output is totalled for each 15 min, as was done by the earlier workers referred to, some fluctuations in production seem to occur (see Fig. 15); but these fluctuations differ widely from day to day and appear to depend on factors totally unrelated to any work rhythm within the operator involved. This seems to be generally true of all situations where people are working under so-called incentive conditions.

Workers operating under non-incentive conditions where a high day rate is paid might still show some form of fluctuation. Only 1 study under these conditions has been recorded; this involved the inspection of pharmaceutical capsules, which is a highly demanding perceptual job. The number and duration of stoppages on this job was recorded over a period of

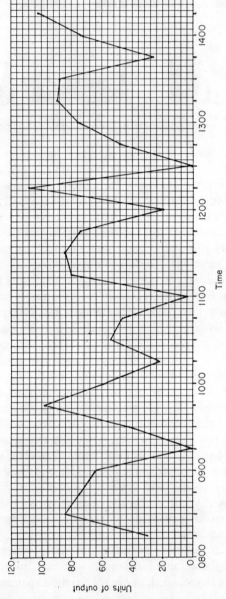

FIG. 15 Quarter-hourly output of the operative in Fig. 14 over the same day.

several weeks. From the records it was possible to calculate the length of time actually worked in each $\frac{1}{4}$ h; a typical result is shown in Fig. 16, which shows that the worker appears to function on cycles which seem to be related to the official breaks in the middle of the morning and the afternoon. However, this must not necessarily be taken as a typical example of industrial performance under modern conditions.

Summing up then, it seems that although rhythms of regularity–irregularity in work which is continuous and monotonous can be demonstrated in the laboratory, modern industrial conditions are such that rhythms of this kind are highly unlikely to occur on the shop floor except perhaps when the work is highly paced. From this it follows that if irregularity is a manifestation of the phenomenon which in the past has been called "fatigue", this fatigue can not now be demonstrated. This is because sufficient rest is either officially permitted or is taken unofficially.

Having said this there is another form of rhythm which does merit consideration. When discussing their work with interviewers administering questionnaires many operatives referred to their dislike of having the "rhythm of their work" interrupted. The "warming-up" phenomenon has already been mentioned; this means in effect that it takes a little while for a worker to speed up to his maximum rate of work and to achieve such a rhythm. However, once he has done so the work is regular—the worker has got into the swing; it is often suggested that interruptions which break the continuity of the work during this period will have an adverse effect on production. The exact extent of this effect is difficult to ascertain. Various investigations into the effect of rest pauses have shown that any disrupting effect is substantially increased if the pause lasts for more than about 10 min, which suggests that when the work rhythm has been established (exactly what this means in neurological terms is open to a wide interpretation) this rhythm may be retained in some way for a period after work has stopped voluntarily. What is not clear is whether it will equally be retained if work is stopped by an **unwanted** interruption. There is some suggestion that the irritation caused by such unwanted interruptions may itself affect the rhythm.

At the present time there are 2 schools of thought on rest requirements in industry. The first and most prevalent of these is that workers should be given the freedom to stop work whenever they want to. This would be reasonable but for 2 factors:

(1) laboratory work has shown that optimum performance is maintained when breaks are taken at times which seem to differ from those at which breaks would be taken voluntarily;

(2) workers do not stop work independently; with girls in particular they tend to "hunt in pairs", e.g., 1 girl decides that she wants to stop

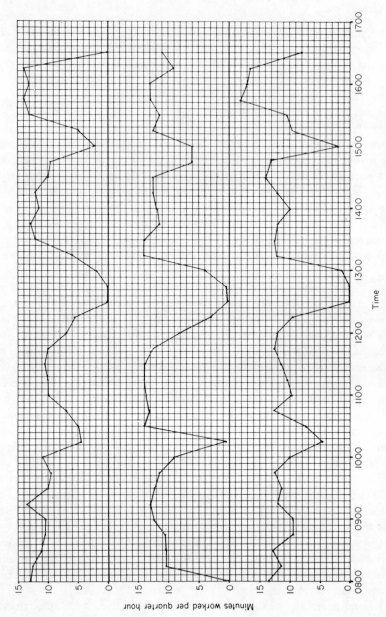

FIG. 16. Mean minutes worked in each 15 min by 7 viewers on 3 successive days.

work, she gets up, goes to a friend, stops her from working and on the way to the toilet the pair may interrupt the work rhythm of a dozen or more other girls. They will do the same thing on the way back. The second view on rest, which is only sparsely held, is that rest should be organized. A recent experiment carried out in the factory of Kalamazoo Limited in Birmingham (Bhatia and Murrell, 1969) showed that there was an increase in production and in earnings when the operatives took a fixed rest period of 10 min in every hour, rather than having the irregularly occurring rests of varying length that they were used to. Questionnaires given during the experiment revealed that one of the factors in the new situation which was liked by the operatives was that they were no longer interrupted during their period of work as they had been previously; this suggests that when they had got into their rhythm of work they preferred to carry on steadily until the time to stop had arrived.

What has just been said is very largely speculative and is based mainly on anecdotal evidence. It is clearly an area which merits further investigation, in particular into the true nature of the warming-up phenomenon and the mechanism by which an operative establishes a rhythm of work which can be maintained for a reasonable period of time.

## III. Between-day Fluctuations

It has already been mentioned that in his experiment on repetitive work Murrell found with his early subjects a marked day-of-the-week effect which caused him to abandon a fully randomized experimental design in favour of administering a standard condition on Tuesdays. The first 5 subjects worked under 3 conditions for $3\frac{1}{2}$ h a day for a varying number of weeks. The sum of the ranks of the daily output for each of these subjects were for the 5 weekdays from Monday to Friday:

$$59, \quad 42, \quad 62, \quad 36\tfrac{1}{2}, \quad 25\tfrac{1}{2}.$$

The overall differences were tested by the Friedman 2-way analysis of variance and were found to be highly significant statistically ($p < 0.001$). Inter-day comparisons showed that Wednesday differed from Tuesday, Thursday and Friday at $p = 0.001$ (Sign test) but did not differ from Monday. Errors for these 5 subjects were also ranked to give sums of:

$$14, \quad 13, \quad 23, \quad 12, \quad 13.$$

The overall differences for these values were also found to be statistically significant by the Friedman test ($p < 0.001$). So it would seem that both in terms of production and errors Wednesday was a poor day. In addition to the data obtained from this experiment, 2 other experiments in which daily scores were available have been traced. In the first (Murrell, 1960) 2 subjects check-read a panel of dials for 2 weeks each. The time taken to notice a deviation was scored. In the second (Mace, 1935) 2 groups of 10 boys were given tests of simple addition under 2 incentive conditions, one group for $3 \times 5$-day weeks, and the other for 2 weeks. Altogether, independent inter-day comparisons are possible for 16 individual subjects or groups of subjects; these have been included in Table II, and show similar trends to the results referred to above. In view of this it seemed worthwhile to investigate whether similar effects could be found in the results of studies carried out in actual industrial situations.

In this, as in other investigations of results obtained in industrial settings, there are a number of intervening variables which make it difficult to assess the true situation in any particular case. The most important of these is the habit, in some companies which are on piece work, for the operatives to keep some production in hand until the final day of the wages week. This follows to some extent from the situation already described where the workers are capable of producing far more than they actually do; when they are working to an earnings target they keep a little bit in hand in case of unforseen circumstances which will prevent them from achieving their target by the end of the wages week. This view is supported by Smith (1953) who says "Even when there is no restriction due to fear of rate-cutting it is normal for any worker to decide in advance how much he will produce, and earn, each day. Effort is unquestionably pegged, at least within narrow ranges in most industrial situations". Other factors which may influence a situation include the incidence of machine break-downs, the accuracy of piece-work standards, or changes in the temperature in the factory, any one of which may affect the results in a random manner.

As with the work already described it proves to be difficult to find much relevant industrial information in the literature, since authors have not generally thought it worthwhile to publish their day-to-day records. So once again we have to go back almost entirely to the work of the Industrial Fatigue Research Board, backed up by some recent work by Murrell and his team. The results can be divided into 2 groups; those which were obtained from actual experiments with industrial subjects, and those which were obtained from records. The data from both sources are given in Table I.

## A. *Experimental studies*

*Study No. 1.* (Smith, 1922): dotting tests were given twice a day to 2 groups of 7 girls in one laundry and to 3 groups of 25, 6 and 4 girls in another for 4 weeks; the number of errors was scored. There was little

TABLE I

*Summary of results from studies of comparative performance on different days of the week, given either as actual scores, or as a ranking of scores, or, where multiple data are available, as sums of ranks.*

| Study No. | Author | | Measure | No. of Ss | M | Tu. | W. | Thu. | F. | Proba-- bility |
|---|---|---|---|---|---|---|---|---|---|---|
| 1 | Smith (1922) | A | Σ ranks | 14 | 14·5 | 9·5 | 14·5 | 8·5 | 13 | — |
| | | B | Σ ranks | 25 | 21 | 25 | 14 | 15 | 16 | — |
| 2 | Murrell and | A | Σ ranks | 2 | 30·5 | 37 | 27 | 31 | 54·5 | <0·01 |
| | Edwards (1963) | B | Σ ranks | 2 | 61 | 64·5 | 84 | 71·5 | 64 | <0·10 |
| 3 | Smith (1922) | rest | rank | 1 | — | 3 | 4 | 2 | 1 | — |
| | | no-rest | rank | 1 | 1 | 4 | 2 | 5 | 3 | — |
| 4 | Weston and Adams (1927) | | Σ ranks | 6 | 18 | 21 | 8 | 23 | 20 | <0·10 |
| 5 | Weston and Adams (1928) | | Σ ranks | 6 | 6 | 9 | 4 | 2 | 9 | <0·20 |
| 6 | Weston and Adams (1928) | | Σ ranks | 5 | 25·5 | 40 | 34 | 23·5 | 27 | <0·20 |
| 7 | Wyatt and Langdon (1938) | | relative output | 34 | 97·4 | 102·1 | 103·6 | 97·9 | 99·0 | — |
| 8 | Wyatt *et al.* (1937) | | hourly output | 68 | 88·5 | 92·1 | 94·7 | 91·5 | 91·6 | — |
| 9 | Wyatt (1923) | | working capacity | — | 7·4 | 5·3 | 8·2 | 8·9 | 10·1 | — |
| 10 | Weston (1922) | | Σ ranks | 3 | 15 | 9 | 4 | 8 | 9 | <0·10 |
| 11 | Farmer and | A | rank | — | 1 | 4 | 5 | 2 | 3 | — |
| | Chambers (1939) | B | rank | — | 5 | 3·5 | 1 | 2 | 3·5 | — |
| | | C | rank | — | 3 | 1 | 4 | 2 | 5 | — |
| | | D | rank | — | 5 | 1 | 4 | 2 | 3 | — |
| 12 | Farmer (1924) | a.m. | Σ ranks | 3 | 12 | 7 | 13 | 6 | 7 | — |
| | | p.m. | Σ ranks | 3 | 9 | 5 | 11 | 5 | 15 | <0·05 |
| | | night | Σ ranks | 3 | 12 | 8 | 8 | 8 | 9 | — |

agreement between the 2 laundries. Girls in laundry A seem to perform well on Tuesday and Thursday but this was mainly due to 1 group and the overall effect is not significant. Thursday is however better than Wednesday in all 4 of the group tests given. In laundry B, Monday and Tuesday appear worse than Wednesday, Thursday and Friday. Although this general trend is not significant Tuesday is worse than Wednesday in all 6 of the tests and is almost significantly different from Wednesday and Thursday ($p = 0.054$).

*Study No. 2.* (Murrell and Edwards, 1963): 4 subjects were tested (2 at a time) on 2 centre lathes, 1 with and 1 without a new measuring device. The task was normal tool room production and the measure used was cutting time per hour. Subjects in the first experiment (A) did well on Wednesday and rather badly on Friday (Wednesday > Friday: $p = 0.003$),

TABLE II

*Number of occasions on which (by the measure used) "performance" was superior on one or other of 2 adjacent days*

| Source of the data | No. of cases | Mon | : | Tues | Tues | : | Wed | Wed | : | Thurs | Thurs: | Fri |
|---|---|---|---|---|---|---|---|---|---|---|---|---|
| Lab. experiments | 16 | 6 | : | 10 | 14 | : | 2 | 1 | : | 15 | 6 : | 8 |
| Ind. experiments | 12 | 8 | : | 2 | 3 | : | 8 | 6 | : | 6 | 5 : | 6 |
| Ind. data with booking | 4 | 0 | : | 4 | 1 | : | 3 | 4 | : | 0 | 2 : | 1 |
| Ind. data without booking | 7 | 1 | : | 6 | 5 | : | 1 | 1 | : | 5 | 6 : | 0 |

but in experiment B the days were reversed (Wednesday < Friday: $p = 0.043$). There was no booking up.

*Study No. 3.* (Smith, 1922): time taken for 1 subject to iron a shirt, with and without rest, on a 9-h day for 4 days a week only. No consistent indication shown.

*Study No. 4.* (Weston and Adams, 1927): 3 subjects linking in the hosiery industry (with and without visual aids) studied over 4 weeks each. Performance measured in time per dozen hose. Booking up is not mentioned. Although the overall day differences just fail to reach significance Wednesday is better than Tuesday and Thursday ($p = 0.008$).

*Study No. 5.* (Weston and Adams, 1928): a study of 6 loomers in the weaving industry, with and without glasses; data given as "relative output". There was booking up.

*Study No. 6.* (Weston and Adams, 1928): a similar experiment with optical aids, using as subjects 2 filament sorters and 3 filament mounters engaged in the manufacture of electric lamps. Both groups of subjects performed worse on Tuesdays and Wednesday than on the other days but the differences are not significant.

B. *Industrial record studies*

*Study No. 7.* (Wyatt and Langdon, 1938): "relative output" for 34 chocolate wrappers studied for 5 weekdays. High performance on Wednesday is said to be due to "booking up".

*Study No. 8.* (Wyatt *et al.*, 1937): the average hourly output of 68 cracker makers was measured over 17 weeks; there was "booking up" on Wednesdays. It is not possible to do statistical tests on the individual subjects' output figures but ranking the days for total output gives a difference with $p < 0.001$ between all days except Thursday and Friday.

*Study No. 9.* (Wyatt, 1923): the daily percentage decrease in "working capacity" in cotton weaving. Booking up is not mentioned. No statistical tests are possible.

*Study No. 10.* (Weston, 1922): 3 different regimes with different numbers of working hours were studied. Booking up took place.

*Study No. 11.* (Farmer and Chambers, 1939): a study of accidents among motor drivers who were divided either according to whether they had 3 accidents in their first year (A) or none (B), or whether they were in the top (C) or bottom (D) quartile on a battery of tests. Data are given as the number of accidents per centum. With the drivers divided in the first manner there are very different weekly patterns, but dividing them by the second criterion gives much more consistent results.

*Study No. 12.* (Farmer 1924): output on 3 shifts in 3 different factories producing glass bottles, recorded over periods covering 6 weeks to $3\frac{1}{2}$ years. Of the data for the 3 individual shifts only those for the afternoon shift show significant differences, with Wednesday and Friday being worse than the other 3 days.

The inter-day differences are summarized in Table II, in which the number of occasions on which the measure used is higher (i.e. performance is better) on adjacent days is given. There is no great agreement between the results obtained from the industrial experiments and those obtained from the collection of industrial data except that overall they both seem to agree that there is unlikely to be a very great difference between Thursdays and Fridays. The industrial experiments suggest that Tuesday is likely to show the lowest performance, which is exactly contrary to the effect found in the laboratory experiments. On the other hand the industrial data suggest that Monday is likely to be the worst day, which is consistent with the popular belief of the existence of a "Black Monday". The results also suggest that there is some truth in the popular idea that the work pattern is influenced by whether there is booking-up or not. When there is booking it will be seen that Wednesday is better than either Tuesday or Thursday; when there is no booking the effect is reversed, with Wednesday being the worst day. The latter finding conforms with the results of the laboratory experiments already mentioned.

It must be admitted that, as with most attempts to look at any form of industrial rhythm, the analysis of the data which have been obtained is rather inconclusive. However, when data are collected over a sufficiently long period of time it seems that there is a probability that a particular group of individuals will establish their own particular rhythm. Where no booking up takes place it seems likely that this will include a comparatively low performance on Mondays and Wednesdays.

# IV. Conclusion

The evidence which has been reviewed in this chapter is both sparse and inconclusive. Whether we are dealing with within-day or between-day fluctuations in performance we have to go to the laboratory to show any consistent trends. This is all very well, but when attempts are made to extrapolate such laboratory findings to the shop floor there are many factors which seem to intervene to prevent a successful outcome. The most important of these would seem to be the system of payment-by-results. Over the day this may result in workers having an excessive amount of time for rest; over the week it may lead to the carry-over of production for booking up on the final day of the wages week. The latter effect can to some extent be overcome by using the method of actual production recording rather than taking as data the output figures as booked by the workers themselves. This is exemplified in an unpublished study by Murrell of performance in a factory where the girls had excessive rest. The booked output showed day-to-day fluctuations, with Monday low and Wednesday high. But when the actual output was recorded it was found to be fairly consistent, with no day differing greatly from another.

So it is that there is no industrial evidence to support the within-day effects obtained in the laboratory. On the other hand the between-day patterns shown by industrial data without booking follow very closely those found in the laboratory experiments. What is perhaps a little surprising is that when experiments are done in industry with industrial subjects the pattern is entirely different. If it is accepted that in spite of the apparent lack of agreement between the shop floor and the laboratory there is some validity in the rhythms which have been demonstrated in the laboratory then a number of management actions can be taken to minimize the effects of what are commonly called "boredom" and "fatigue" and to increase work satisfaction. So this work, inconclusive though it is, can have some practical outcome.

## References

Arai, T. (1912). "Mental Fatigue". Contributions to Education, No. 54. Teachers College, Columbia University.

Baker, C. H. (1959). *Canad. J. Psychol.* **13**, 35–42.

Baldamus, W. (1951). "Incentives and Work Analysis". University of Birmingham. Studies in Economics and Society, Monograph A.

Bartlett, F. C. (1953). *In* "Symposium on Fatigue" (W. F. Floyd and A. T. Welford, eds), pp. 1–5. H. K. Lewis, London.

Bergum, B. O. (1966). *Percept. Mot. Skills* **23**, 47–54.

Bhatia, N. and Murrell, K. F. H. (1969). *Hum. Fact.* **11**, 167–174.

Bills, A. G. (1931). *Amer. J. Psychol.* **43**, 230–245.

Broadbent, D. E. and Gregory, M. (1963). *Brit. J. Psychol.* **54,** 309–323.
Burnett, I. (1924). *J. Nat. Inst. Ind. Psychol.* **2,** 18–23.
Burtt, H. E. (1929). "Psychology and Industrial Efficiency". Appleton, New York.
Farmer, E. (1924). "A Comparison of Different Shift Systems in the Glass Trade". I.F.R.B. Report 24. H.M.S.O., London.
Farmer, E. and Chambers, E. G. (1939). "A Study of Accident Proneness among Motor Drivers. I.H.R.B. Report 84. H.M.S.O., London.
French, J. D., Hernández-Péon, R. and Livingston, R. B. (1955). *J. Neurophysiol.* **18,** 74–95.
Hersey, R. B. (1925). *J. Personn. Res.* **4,** 37–45.
Huxtable, Z. L., White, M. H. and McCartor, M. A. (1946). *Psychol. Monogr.* **59,** No. 275.
Jerison, H. J. (1967). *In* "Attention and Performance" (A. F. Sanders, ed.), pp. 373–389. North-Holland, Amsterdam.
Katzell, R. A. (1950). *In* "Handbook of Applied Psychology" (D. H. Friar and E. R. Henry, eds), pp. 74–84. Holt, Rinehard and Winston, New York.
Mace, C. A. (1935). "Incentives and some Experimental Studies". I.F.R.B. Report 72. H.M.S.O., London
Manzer, C. W. (1927). *Arch. Psychol.* **90.**
Murrell, K. F. H. (1960). *Ergonomics* **3,** 213–244.
Murrell, K. F. H. (1962). *Int. J. Prod. Res.* **1,** 39–55.
Murrell, K. F. H. (1965). *Bull. Cent. Etud. Rech. Psychotech.* **14,** 104–110.
Murrell, K. F. H. (1966). *In* "Proceedings of the Second Seminar on Continuous Work" (F. F. Leopold, ed.), pp. 122–126. Institute for Perception Research, Eindhoven.
Murrell, K. F. H. (1967). *In* "Attention and Performance" (A. F. Sanders, ed.), pp. 427–435. North-Holland, Amsterdam.
Murrell, K. F. H. (1969). *Acta Psychol.* **29,** 268–278.
Murrell, K. F. H. and Edwards, E. (1963). *Occup. Psychol.* **37,** 267–275.
Murrell, K. F. H. and Forsaith, B. (1963). *Int. J. Prod. Res.* **2,** 247–264.
Phillpott, O. O. (1932). *Brit. J. Psychol. Monogr. Suppl.* **6,** 1–125.
Poffenberger, A. T. (1928). *J. Appl. Psychol.* **12,** 459–464.
Reed, H. B. (1924). *J. Educ. Psychol.* **15,** 389–392.
Robinson, E. S. and Heron, W. T. (1924). *J. Exp. Psychol.* **7,** 81–97.
Roethlisberger, F. J. and Dickson, W. J. (1941). "Management and the Worker". Harvard Univ. Press, Cambridge, Mass.
Rothe, H. F. (1946). *J. Appl. Psychol.* **30,** 199–211.
Siegel, L. (1962). "Industrial Psychology". Homewood, Irwin.
Smith, M. (1922). "Some Studies in the Laundry Trade". I.F.R.B. Report 22. H.M.S.O., London.
Smith, P. C. (1953). *J. Appl. Psychol.* **37,** 69–74.
Vernon, H. M. (1924). *Brit. J. Psychol.* **15,** 393–404.
Vernon, H. M., Bedford, T. and Wyatt, S. (1924). "Two Studies on Rest Pauses in Industry". I.F.R.B. Report 25. H.M.S.O., London.
Weiland, J. D. (1927) *Arch. Psychol.* **87.**
Weston, H. C. (1922). "A Study of Efficiency in Fine Linen Weaving" (Textile Series No. 5). I.F.R.B. Report 20. H.M.S.O., London.
Weston, H. C. and Adams, S. (1927). "The Effect of Eyestrain on Output of Liners in the Hosiery Industry". I.F.R.B. Report 40. H.M.S.O., London.

Weston, H. C. and Adams, S. (1928). "On the Relief of Eyestrain among Persons Performing Very Fine Work". I.F.R.B. Report 49. H.M.S.O., London.

Wyatt, S. (1923). "Variations in Efficiency in Cotton Weaving" (Textile Series No. 7). I.F.R.B. Report 23. H.M.S.O., London.

Wyatt, S. (1929). *Personnel J.* **8**, 161–171.

Wyatt, S. and Fraser, J. A. (1925). I.F.R.B. Report 32. H.M.S.O., London.

Wyatt, S., Langdon, J. N. and Stock, F. G. L. (1937). "Fatigue and Boredom in Repetitive Work". I.F.R.B. Report 77. H.M.S.O., London.

Wyatt, S. and Langdon, J. N. (1938). "The Machine and the Worker. A Study of Machine Feeding Process". I.F.R.B. Report 82. H.M.S.O., London.

# Author Index

*Numbers in italics refer to pages on which a reference is listed at the end of a chapter*

## A

Aarons, L., 52, *105*
Abramson, M., 213, 214, *239*
Abt, L. E., 162, *174*
Adams, O. S., 62, 64, *105*
Adams, S., 267, 268, *271*
Adams, T., 59, *106*
Adkins, S., 53, 64, *105*
Agnew, H. W., Jr., 150, 154, 155, 157, 158, 159, 165, 166, 167, *174*, *176*, *177*
Akert, K., 14, *36*, *37*
Allport, D. A., 199, *207*
Alluisi, E. A., 53, 62, 64, *105*
Alpern, M., 21, *36*
Altmann, M., 213, 236, *239*
Anon, 216, *239*
Arai, T., 242, *270*
Arey, L. B., 232, *239*
Armington, J. C., 201, *209*
Aschoff, J., 7, *36*, 58, 60, *105*, 171, *174*
Aserinsky, E., 150, *174*
Axelrod, J., 34, *38*

## B

Baddeley, A. D., 48, *105*, 188, *207*
Baekeland, F., 154, *174*
Baker, C. H., 259, *270*
Balaban, M., 10, *36*
Balasz, J. V., 215, *239*
Baldamus, W., 245, *270*
Bartlett, F. C., 255, *270*
Bates, J. A., 192, *207*
Beck, E. L., 190, *207*
Bedford, T., 243, *271*
Bekesy, G. von, 30, *36*
Bell, C. R., 134, *147*
Bendict, F. G. 58 *105*, 213, *239*
Benedek, T., 213, 233, *239*
Beraud, G., 156, *176*
Bergum, B. O., 259, *270*

Bernhard, C. G., 190, *207*
Betz, A., 9, *36*
Bhatia, N., 265, *270*
Bickers, W., 215, *239*
Bills, A. G., 242, *270*
Bjerner, B., 55, *105*
Blackman, R. B., 28, *36*
Blake, M. J. F., 48, 49, 68, 72, 73, 76, 83, 86, *105*, 111, 112, 113, *147*
Bockh, H., 52, *105*
Bogoslovsky, A. I., 52, *105*
Bost, J., 156, *176*
Bradbury, P. A., 48, *105*
Bramson, J., 212, *239*
Brandis, S., 213, 216, 228, *239*
Bremer, F., 147, *147*
Broadbent, D. E., 13, *36*, 198, *207*, 259, *270*
Brown, I. D., 52, *105*
Browne, R. C., 55, 56, *105*
Bruce, J., 213, 215, *239*
Bruce V. G., 8, *36*
Brunelli, F., 217, 228, 236, *239*
Bruner, H., 59, 64, *106*
Bull, H. D., 213, 236, *239*
Bunning, E., 7, *36*
Burnett, I., 256, *271*
Burton, A. C., 59, *105*
Burtt, H. E., 244, 247, 249, *271*

## C

Callaway, E., III, 190, *207*
Cannon, L. D., 53, *105*
Carlson, V. R., 155, 169, *175*
Chambers, E. G., 267, 269, *271*
Chance, B., 9, *36*
Chang, H. T., 196, 206, *207*
Chapanis, A., 201, 203, *207*
Cheatham, P. G., 201, 202, *207*, *209*
Chiles, W. D., 53, 62, 64, *105*
Clemes, S. R., 158, *174*
Cloudsley-Thompson, J. L., 172, *174*

Kripke, D. F., 156, 166, *175, 177*
Kristofferson, A. B., 185, 186, 187, 188, *208*
Kroger, W. S., 215, *240*
Kropfl, W., 192, 193, *207*
Kun, T., 157, *175*

**L**

Ladefoged, P., 198, *207*
Lamb, W. M., 214, *240*
Landauer, T. K., 205, *208*
Langdon, J. N., 248, 249, 250, 253, 267, 268, *272*
Lansing, R. W., 189, 192, *208*
Lashley, K. S., 14, *37*
Lasky, R., 154, *174*
Lategola, M. T., 53, *105*
Latour, P. L., 20, *37*, 196, *208*
Lederer, J., 215, *240*
Legg, C. F., 48, *105*, 188, *208*
Le Grand, Y., 203, *208*
Lehmann, G., 49, *106*
Leonard, J. A., 122, *147*
Lesèvre, N., 182, *207*
Levitt, R. A., 167, *175*
Lewin, K., 215, 228, *240*
Lewis, H. B., 154, *177*
Lewis, O. F., 156, *175*
Lhamon, W. T., 48, *106*
Lichtenstein, M., 199, 201, 203, 204, *208*
Licklider, J. C. R., 21, 30, *37*
Lipscomb, H. S., 66, 67, *105*
Livingston, R. B., 259, *271*
Lobban, M. C., 10, *37*, 58, *106*
Lockhart, A., 218, *240*
Lockhart, J. M., *106*
Loeser, A. S., 232, *240*
Loomis A. L. 150, *175, 176*
Loon, J. H., van, 58, *106*
Lough, O. M., 216, *240*
Loveland, N. T., 52, *106*
Lubin, A., 51, 52, *106*, 137, *148*
Luce, G. G., 162, 176
Luchian, O., 57, *106*
Luff, M. C., 214, 236, *240*

**M**

McCance, R. A., 214, 236, *240*
McCartor, M. A., 242, *271*

McCulloch, W. S., 180, *208*
McGhie, A., 169, *176*
McGill, W. J., 195, *208*
McGreggor, P., 166, *177*
Mackinnon, I. L., 212, 213, *240*
McReynolds, P., 181, *208*
Mackworth, N. H., 122, *147*
Mace, C. A., 266, *271*
Mallay, H., 162, 164, 169, *176*
Manzer, C. W., 242, *271*
Mare, G. de la, 57, *106*
Marriott, R., 57, *107*, 166, *177*
Marshall, J., 213, 232, 234, *240*
Melton, A. W., 146, *147*
Menaker, M., 212, *240*
Menzel, W., 57, *106*
Merrill, M. A., 132, *147*
Merton, P. A., 20, *36, 37*
Metz, B., 68, *107*
Michel, F., 156, *175*
Middleton, W. C., 215, *240*
Mihaila, I., 57, *106*
Milhorn, T., 30, *37*
Miller, G. A. 205, *208*
Mills, J. N., 16, 34, *37*, 40, *106*
Minorsky, N., 2, 11, *37*
Mitchell, M. B., 213, *240*
Mohler, S. R., 32, *37*
Moore, L. M., 212, *240*
Moore, T., 168 *176*
Morton H. B. 20 *37*
Morton J. H., 213, 215, 216, *240*
Mundy-Castle, A. C., 182, *208*
Murphree, O. D., 201, *208*
Murray, E. J., 51, 52, *106*, 162, *176*
Murrell, K. F. H., 56, *106*, 250, 254, 255, 256, 257, 258, 259, 260, 265, 266, 267, *270, 271*
Muzio, J. N., 155, *176*
Myers, C. S., 217, 218, 220, 222, 232, 234, *240*

**N**

Neisser, U., 13, *37*
Newman, E. A., 17, *37*
Nogeire, C., 166, *177*

**O**

Oatley, K., 21, 33, *37*
Obrist, W. D., 185, *208*

# Subject Index